T0324750

The Auchenorrhyncha
(Homoptera)
of Fennoscandia and Denmark

FAUNA ENTOMOLOGICA SCANDINAVICA

Volume 7, part 1   1978

The Auchenorrhyncha (Homoptera) of Fennoscandia and Denmark

Part 1: Introduction, infraorder Fulgoromorpha

by

F. Ossiannilsson

SCANDINAVIAN SCIENCE PRESS LTD.

Klampenborg . Denmark

© *Copyright*
Scandinavian Science Press Ltd. 1978

Fauna entomologica scandinavica
is edited by »Societas entomologica scandinavica«

Managing editor
Leif Lyneborg

World list abbreviation
Fauna ent. scand.

Text composed and printed by
Vinderup Bogtrykkeri A/S
7830 Vinderup, Denmark

Plates reproduced and printed by
Grafodan Offset
3500 Værløse, Denmark

Publication date
1 October, 1978

ISBN 87-87491-24-9

Editors preface

Volume 7 of "Fauna entomologica scandinavica" will treat the suborder Auchenorrhyncha of the order Homoptera. The auchenorrhynchous Homoptera is a large group with about 415 species in Denmark and Fennoscandia.

Volume 7 will appear in three parts. Part 1 comprises the introductory chapters and the infraorder Fulgoromorpha with 88 species in 4 families (Cixiidae, Delphacidae, Achilidae and Issidae). Parts 2 and 3 will deal with the infraorder Cicadomorpha including 327 species.

It may expected that part 2, comprising the families Cicadidae, Cercopidae, Membracidae and Cicadellidae but excluding the subfamily Deltocephalinae, will appear in 1980. Part 3 with the subfamily Deltocephalinae will probably be published during 1982. This part deals with 145 species and will also include the catalogue and the list of references.

Parts 1-3 of volume 7 will be provided with continous numbers for both pages and figures.

LEIF LYNEBORG

Introduction

Our knowledge of the Auchenorrhyncha of Fennoscandia and Denmark started with Linnaeus and his work on the Swedish fauna. Following Linnaeus studies were continued largely on a local or country by country basis until the present. Thus our review of past work starts with that of Sweden, followed by those of Norway, Denmark and Finland.

Of the 24 insects described by Linnaeus as *Cicada* species in "Fauna Svecica" (1761), one (*aptera*) belongs to Heteroptera, one (*flava*) is of uncertain systematic position, and four are synonyms. Charles de Geer, a contemporary of Linnaeus, was especially important as a local observer and describer of insect life, very much in advance of his time. He also described a number of species. Other important contributors to our knowledge of Swedish Homoptera were C. F. Fallén (1805, 1806, 1814, 1826) and J. W. Zetterstedt (1828, 1838). All auchenorrhynchous Homoptera described as new species in Zetterstedt's "Insecta Lapponica" except *Cicada lineigera*, must be dated 1838, not 1840 as is usually done. Further additions to the auchenorrhynchous fauna of Sweden, including many new species, were made by A. F. Dahlbom (1850), C. H. Boheman (1845-1867), C. Stål (1853-1878), C. G. Thomson (1869, 1870), and H. D. J. Wallengren (1851, 1870). Stål was especially important as he was a taxonomic reviser on a world basis. I started my studies on the Swedish fauna in 1932. A key to the Swedish species appeared in 1946-47 and a tabular catalogue in 1948. The enthusiastic work of N. Gyllensvärd (1961-1972) has considerably increased the known fauna of Sweden. His work has filled many gaps in our catalogue.

During his travels in "Lapponia" Zetterstedt also visited Norway and made the first contributions to the Norwegian fauna in his books (vide supra). This fauna was later more directly covered by Siebke (1870, 1874). J. Sahlberg (1871) monographed the auchenorrhynchous Homoptera of Finland, Sweden and Norway. Sahlberg's species descriptions are much more detailed than those of earlier authors and he also presented a tabular catalogue showing species distribution in the three countries. In a later paper Sahlberg (1881) contributed to the knowledge of the fauna of·Northern Norway. W. M. Schøyen (1879) supplemented Siebke's work. Embrik Strand (1902, 1905, 1913) recorded a number of finds in Norway but his work is perhaps not always quite trustworthy. Some of my papers (1943c, 1962, 1974, 1977) are devoted to the Norwegian fauna. H. Holgersen also has published on this fauna in a valuable series of papers (1944-1954). Unfortunately only parts of the group were dealt with in this series but we are hoping for a continuation.

The fundamental publications on the Danish auchenorrhynchous fauna are by O. Jacobsen (1915) and A. C. Jensen-Haarup (1915-1920). Some additions and corrections to the Danish list were published by W. Wagner (1935). Also valuable contributions were made by N. P. Kristensen (1965a, 1965b) and L. Trolle (1966, 1968, 1973, 1974).

An outstanding contributor to the knowledge of the Finnish Auchenorrhyncha fauna is Håkan Lindberg (1923–1952). He also contributed to the Norwegian and Swedish faunas. His studies on the stylopization of *Chloriona* and other Delphacidae (1939, 1949) are very important. In 1947 he published an annotated list of the Auchenorrhyncha of East Fennoscandia. C. R. Sahlberg (1842) described three new species from Finland. Kontkanen (1937–1954) and Linnavuori (1948–   ) studied the Auchenorrhyncha of Finland both faunistically and ecologically. Linnavuori's treatise on the Auchenorrhyncha in Animalia Fennica 12–13 (1969) is very generously illustrated and serves as a good aid to the identification of Scandinavian species despite the text being entirely in Finnish. M. Raatikainen and his collaborators (1959–   ), while studying the biology of planthoppers and leafhoppers mainly in their role as pests of cultivated plants, have also produced very valuable information on their distribution and ecology. Recently Huldén (1971–   ) has started promising studies on the distribution of Finnish Hemipteroidea.

As in many insect groups, the generic and specific taxonomy of Auchenorrhyncha is based on male genital structure. Originally only the parts externally visible were studied. Fieber (1869) used the shape of the styles as a diagnostic character in the "Deltocephali". From 1878 onward, J. Edwards used successfully the structure of aedeagus and styles as species characters. Also Then (1896 and later) did likewise. However the use of male genitalia for species separation was controversial; many entomologists doubted the validity of species separated as such. As late as 1935, H. Haupt (pp. 217–218) expressed such doubts. Nowadays nobody agrees with him. Geographic variation in the shape of the appendages of the aedeagus exists, e. g. in *Philaenus spumarius* (W. Wagner, 1955a). The hibernating generation of *Euscelis incisus* differs considerably from summer generations in the shape of the aedeagus (Müller, 1947 and later). These differences were originally regarded as specific but Müller showed that they were caused by different day-length.

Ribaut (1942a and b, 1946) revised the Deltocephalinae, basing his taxonomy not only on male genitalia but also on the chaetotaxy of legs and genital plates. W. Wagner (1962) made a phylogenetic system for the Central European Delphacidae. This resulted in the erection of numerous new genera. As Wagner based his system on tendencies rather than on well-defined morphological characters, the construction of a key to the genera of the family was not an easy matter. Therefore the key to this group is partly based on more or less superficial characters. Dlabola (1958) reclassified the palaearctic Typhlocybinae. Further work in this field has been done by G. A. Anufriev and I. Dworakowska. These revisions are summarised in the check list by J. Nast (1972). A generic revision of palaearctic Idiocerinae was presented by Dlabola (1974).

The colour illustrations for the present book were prepared by Mrs Grete Lyneborg. Most of the pencil-drawings made by myself for "Svensk insektfauna" (Ossiannilsson, 1946–47) have been used as plate-figures also in this book. The ink drawings were also prepared by myself, in most cases directly after nature. A few illustrations by others have been copied. The male genitalia of almost all species have been figured. Only some very rare species, such as *Cixidia lapponica*, have not been figured. The genitalia of females are figured only when they offer useful characters. In the Delphacidae the shape of the genital scale of females is useful as species character. Previously this character has only been used in *Delphax* (Ribaut, 1934, Linnavuori, 1955).

Synonyms are given only if they have been used in older Danish or Fennoscandian literature. Names of colour varieties are not considered.

For each species the distribution in and outside of Denmark and Fennoscandia is briefly given. The information concerning Denmark was supplied by Mr Lars Trolle, for Finland by Mr Larry Huldén, and for Norway and Sweden from my own records. The latter includes records from identifications done by myself for museums and private collections over the last 45 years as well as those from the literature. Distribution data for outside of Fennoscandia and Denmark were mainly derived from Nast (1972). Professor Dr Reinhard Remane kindly checked the data for Northern Germany, the southern boundary for which is 53°N latitude.

Information on the biology of the species has been compiled from the literature or my own observations. The literature cited list was made as complete as possible.

Acknowledgements

For the kind loan of material from various institutions I am much indebted to Dr Astrid Løken, Zoological Museum, Bergen; Director Holger Holgersen, Stavanger Museum; Dr Martin Meinander, Zoological Museum, Helsingfors; Dr Juhan Vilbaste, Zoological and Botanical Institure, Tartu; Professor Dr Lars Brundin, Professor Dr Edvard Sylvén, and Mr Per Inge Persson, Natural History Museum, Stockholm; Professor Dr Carl H. Lindroth, Professor Dr Bengt-Olof Landin, Dr Hugo Andersson and Mr Roy Danielsson, Zoological Museum, Lund; and Dr Henrik W. Waldén, Natural History Museum, Göteborg. Mr Holgersen also lent me great parts of his private collection. Dr W. J. Le Quesne, Chesham, Bucks., England, lent me specimens out of his collection for reproduction of genitalia. Professor Dr Janusz Nast, Institute of Zoology, Warszawa; Dr Irena Dworakowska, Warszawa; and Mr Lars Trolle, Østermarie, Denmark, kindly presented me with specimens of certain rare species. For gifts of valuable specimens I also thank Mr Tord Tjeder, Rättvik, Sweden; Mr Bengt Ehnström, Stockholm; and Mr Bo Henriksson, Edsbyn, Sweden.

I am especially indebted to Professor Dr. Reinhard Remane, Zoological Institute of the Philipps-Universität, Marburg/Lahn, BRD; Mr Lars Trolle, Østermarie; and Mr Larry Huldén, Zoological Museum of Helsingfors, Finland, for their kind help with German, Danish, and East Fennoscandian parts of the catalogue, and for much information.

I thank Mrs Grete Lyneborg for her excellent colour illustrations, and Dr Leif Lyneborg for directions, advice and help in preparing the manuscript.

Last, but not least, I thank my friend, Dr Nils Gyllensvärd, Karlskrona, for partly supporting the printing costs. Also I have valued his stimulating cooperation over the many years.

Survey of the superorder Hemipteroidea

The Auchenorrhyncha are here treated as a suborder of the order Homoptera. Together with the Heteroptera, the Homoptera constitute the superorder Hemipteroidea (= Hemiptera or Rhynchota in older literature). As these two orders are closely related, I will consider the entire superorder.

The Hemipteroidea are hemimetabolous insects. Their mouthparts are piercing and suctorial, consisting of maxillae and mandibles modified and prolonged into two pairs of stylets, which at rest are more or less retracted in the head, with their apical parts being enclosed in the grooved, usually segmented, labium or rostrum. The fore wings are usually larger or at least longer than the hind wings. During flight the wings are usually held together by means of a wing-coupling apparatus. In most groups the anal field of the fore wings is demarcated from the rest of the wing by a straight suture, the claval suture. The presence of this so-called clavus in both orders is, like the common structure of the mouthparts, strong evidence for a close (sister-group) relationship of the Homoptera and the Heteroptera. The wings may be more or less reduced.

The external body structure of the Hemipteroidea is variable. Among the common features mentioned above, the structure of the mouthparts deserves special attention. The stylets are mutually free at base but unite at the apex of the hypopharynx. Each maxillary stylet is grooved along its inner surface, the groove being divided into two channels by a longitudinal ridge. The two maxillary stylets are closely interlocked, or at least opposed to one another, resulting in the formation of two longitudinal canals, a dorsal suction canal, and a ventral ejection canal for the saliva. Laterally the maxillary stylets are flanked by the mandibular ones. The latter are apically armed with barblike sawteeth and can be moved by special muscles. The labium is transformed into a grooved, at most 4-segmented sheath or rostrum, which at rest encloses the apical parts of the stylets. The labial groove is closed along most of its length by its margin being contiguous basally where the gap is covered by the short labrum. In some groups the stylets are very long, at rest being carried rolled-up either in a pocket in the head, or sometimes also in the prothorax, or, as in preadult instars of psyllids, outside the body. By various interlocking arrangements the stylets can move along with each others without loosing contact. By aid of the barbs of the mandibulary stylets these can be forced deeper and deeper into the substrate by alternating movements. The labium does not take part in this penetration; usually it is knee-bent or compressed like a concertina during the act.

Wings may be normal or more or less reduced. In Homoptera as well as in Heteroptera many species are wing-polymorphous, i. e. in one and the same species

9

there are individuals with normal wings and others with more or less reduced wings. If there are only two alternatives in a species this is said to be wing-dimorphous. Often the wing-polymorphism is related to the sex. In the Coccoidea the females are always wingless, while males may be either winged or wingless. In winged coccid males the hind wings are reduced into a pair of hook-like appendages. In aphids there is a more or less regular alternation of winged and wingless generations in connection with alternation of hostplants.

Reproduction is usually normal (amphigony). In some coccids, psyllids, aleyrodids and Heteroptera, constant or, in others, facultative parthenogenesis is found, while cyclic parthenogenesis regularly alternating with amphigony is common among aphids. Ovoviviparity occurs e. g. in certain coccids, viviparity among aphids and some coccids and Heteroptera. Homosexual behaviour has been identified in the males of certain Heteroptera.

Development is gradual. In male coccids, however, there is a pupa-like instar. In aleyrodids the development of wing buds is postponed to the last preimaginal instar.

All Hemipteroidea live on fluids, mostly of vegetable origin. Among the Heteroptera there are also predators and animal parasites. Many species are strongly specialized in their choice of nutrition, others less fastidious. Some species of Heteroptera attack both plants and animals. A regular host alternation is more or less obligatory in many aphid species. Characteristic deformations or even galls are produced by the attack of certain Hemipteroidea on their hostplants. Many Homoptera are vectors for plant diseases caused by virus or mycoplasma. Endosymbiosis with bacteria or yeasts is general in the Hemipteroidea.

Most Hemipteroidea are terrestrial insects. However, some Heteroptera live in fresh or brackish water, others crawl or jump on the surface of water. Most Auchenorrhyncha, Psylloidea and some Heteroptera and aphids are capable of leaping. While most Heteroptera, Auchenorrhyncha and adult Psylloidea are more or less mobile and active insects, aphids and nymphs of psyllids and aleyrodids are more or less sedentary, and most adult coccid females have reduced legs and are incapable of moving from the site to which they have attached themselves.

Stridulation occurs in some Heteroptera, aphids and psyllids. A sound-producing apparatus of quite special construction is the tymbal organ of the Auchenorrhyncha.

Classification of Hemipteroidea

1	Head with a distinct sclerotized "throat" (gula)	**Heteroptera**
–	Head with gula membranous or wanting	**(Homoptera)** 2
2 (1)	Rostrum distinctly arising from the head	3
–	Rostrum apparently arising between or behind anterior coxae. Tarsi 1- or 2-segmented	**(Sternorrhyncha)** 4

3 (2) Rostrum basally flanked by propleural parts   **Coleorrhyncha**
–   Rostrum basally free   **Auchenorrhyncha**
4 (2) Tarsi 1-segmented. Adult females always wingless, often sedentary with reduced legs, males with at most one pair of usable wings   **Coccoidea**
–   Tarsi 2-segmented. 2 pairs of wings present in winged forms   5
5 (4) Clavus absent. Antennae at most 6-segmented. Often wingless   **Aphidoidea**
–   Clavus present. Always winged in adults   6
6 (5) Wings without a mealy wax coating. Wing-coupling apparatus present. Ocelli 3. Antennae 8-, 9-, or 10-segmented   **Psylloidea**
–   Wings covered by a mealy white wax powder. Wing-coupling apparatus absent. Ocelli 2. Antennae 7-segmented   **Aleyrodoidea**

Morphology and diagnostic characters of the Auchenorrhyncha

The following account does not give a complete morphological description of the Auchenorrhyncha but only explains the terminology used in our work.

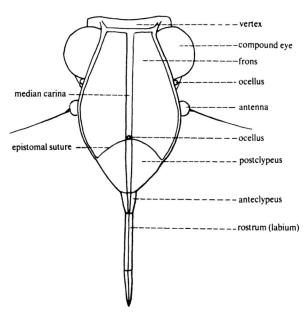

Text-fig. 1. *Cixius similis* Kirschbaum (Cixiidae), face.

Head (Text-figs. 1–6)

The compound eyes are situated laterally on the head. The vertex is the posterior part of the upper region of the head. The coronal suture, a longitudinal suture, divides the vertex into two parts. The frons lies in front of the vertex. The genae and antennae lie lateral to the frons and the clypeus behind (but in morphological sense in front of) the frons. The frons is limited by the following sutures: the epicranial suture on the vertex, the frontal sutures on the genae, and the epistomal suture on the clypeus. The border of the genae on the vertex is usually indistinct. Anteriorly the genae run into the maxillar plates, with or without a distinct boundary-line. A transverse furrow divides the clypeus into two parts, an upper (posterior) part, postclypeus, and a lower (anterior) part, anteclypeus. The lora or mandibular plates are delimited lateral lobes of the clypeus. On each side between the lorum and the maxillar plate runs the genal suture. The true clypeus is laterally demarcated by the clypeal sutures. However, all these sutures are rarely coexisting. Thus, the epicranial suture is usually missing. The epistomal suture is absent in Membracidae, most Cicadellidae and some Cercopidae, frons and postclypeus coalescing into one piece, the frontoclypeus. If present, the ocelli are either three or two in number. The antennae are filiform with cylindrical segments in the Cicadidae. In the other Auchenorrhyncha the two basal antennal segments are much larger and thicker than the others, the latter form together a thong-shaped flagellum (Text-figs. 4 and 5). The flagellum is apically thickened into a so-called

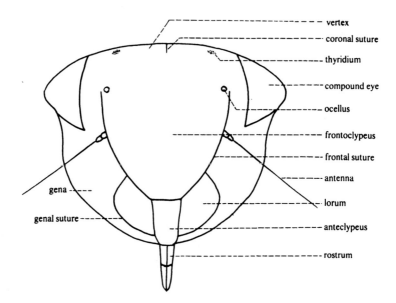

Text-fig. 2. *Hesium domino* (Reuter) (Cicadellidae), face.

12

palette in the males of some Idiocerinae (Text-fig. 6). In front of the anteclypeus lie the labrum and the rostrum. In Cicadellidae a pair of small roundish spots differing from the environment in surface structure, the so-called thyridia, can often be observed near the posterior border of vertex (see Text-fig. 2). The parts of the integument of the postclypeus (frontoclypeus), to which the musculature of the suction pump is attached, are often marked on the outside by a different surface structure and by a pigmentation differing from the rest. In the present book such markings are called muscle traces.

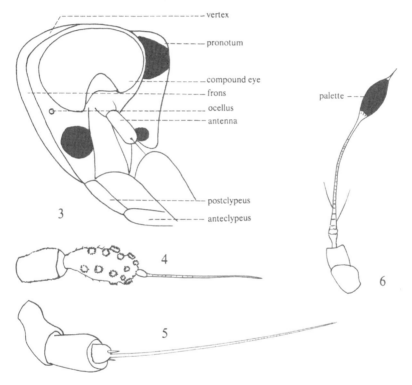

Text-fig. 3. *Kelisia vittipennis* (J. Sahlberg) (Delphacidae), head and thorax from the left.
Text-fig. 4. *Javesella obscurella* (Boheman) (Delphacidae), antenna.
Text-fig. 5. *Lepyronia coleoptrata* (Linné) (Cercopidae), antenna.
Text-fig. 6. *Populicerus populi* (Linné) (Idiocerinae), antenna.

Thorax

The prothorax, the foremost segment, is usually fairly short. The pronotum, its tergal

13

part, is a transverse plate. The mesothorax, the second thoracic segment, carries the fore wings. The mesonotum, the tergum of the mesothorax, consists of four more or less distinct parts arranged in order from the front: prescutum, scutum, scutellum, and postscutellum. In this book the part of mesonotum visible from above in specimens with wings in a resting position is called "scutellum". This is practised in spite of its different composition in the various groups: in Fulgoromorpha and Cicadidae the scutellum consists of the major part of mesonotum, in Cercopidae of the central part of the true scutellum, in Cicadellidae of the caudal part of scutum + the middle part of scutellum. A small scale-like plate, the tegula, covers the base of the fore wing in the Fulgoromorpha (see Text-fig. 10). Sometimes the tegulae are concealed under the hind border of prothorax. Metathorax, the third thoracal segment, carries the hind wings, and metanotum, its tergal part, is entirely concealed by the wings in repose.

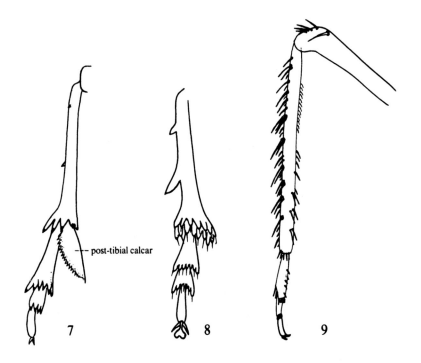

Text-fig. 7. *Javesella obscurella* (Boheman) (Delphacidae), right hind tibia from below.
Text-fig. 8. *Lepyronia coleoptrata* (Linné) (Cercopidae), hind tibia and tarsus from below.
Text-fig. 9. *Scleroracus russeolus* (Fallén) (Cicadellidae), apex of hind femur, tibia and tarsus.

Legs (Text-figs. 7–9)

These consist of the same elements as those of other insects. The tibiae are often armed with fixed spines and/or movable setae, the latter usually arranged in longitudinal rows. The hind femora in the Cicadellidae are apically armed with a few strong setae, the number of which is constant within the various taxa, giving a character useful in keys. Also the chaetation of the dorsal surface of the anterior and median tibiae furnishes

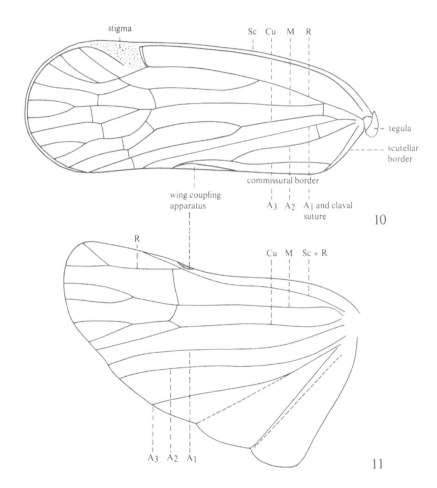

Text-fig. 10. *Cixius cunicularius* (Linné) (Cixiidae), fore wing and tegula.
Text-fig. 11. Same, hind wing.

characters for the separation of genera, especially in the Deltocephalinae. Characteristic of the Delphacidae is the so-called post-tibial calcar (Text-fig. 7), which varies within the family in shape, size, and number of marginal teeth as well as in the degree of development of its apical tooth.

Wings (Text-figs. 10–19)

The fore wings may be leathery or membranous. Even in the latter case they are usually somewhat firmer than the hind wings. In many species there is a polymorphism in the development of the wings. The wing-polymorphism is usually a dimorphism, i. e. there are two different forms, but also trimorphous species with 3 different forms do exist. In

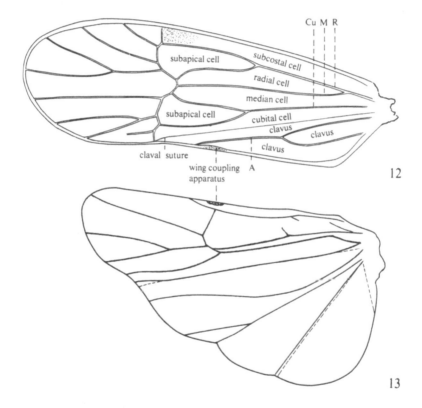

Text-fig. 12. *Javesella pellucida* (Fabricius) (Delphacidae), fore wing of macropterous specimen.
Text-fig. 13. Same, hind wing of macropterous specimen.

a typical wing-dimorphous species there are a short-winged (brachypterous) form with more or less shortened, usually leathery, fore wings covering the reduced hind wings unusable for flying, and a long-winged (macropterous) form with normal wings, the fore wings being either leathery or membranous. In the trimorphous species there is also a third (intermediary) form. Sometimes wing-dimorphism is more or less completely correlated with sex, most males for instance being macropterous while the females are largely brachypterous.

The longitudinal veins in fore and hind wings of macropterous specimens are here denominated in accordance with the generally applied system: costa (C), subcosta (Sc), radius (R), media (M), cubitus 1 (Cu_1), cubitus 2 (Cu_2), and anales (A). In the various groups these veins are more or less well developed and ramified, or reduced. A few examples are illustrated in Text-figs. 10–19.

The true costa is present as an independent vein in the fore wing only in a few families on a low anagenetic level. In the others the costa is reduced or fused with the subcosta, the latter replacing the costa in location and function. The spaces between veins are called "cells". The cell between Sc and R is the subcostal cell, the one between R and M the radial cell, the one between M and Cu the median cell, and the one

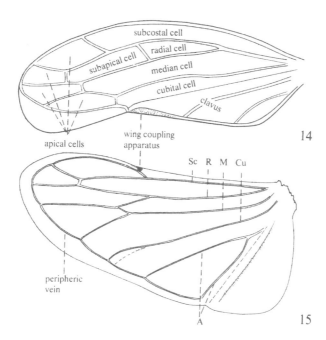

Text-fig. 14. *Populicerus populi* (Linné) (Idiocerinae), fore wing.
Text-fig. 15. Same, hind wing.

between Cu and the claval suture is the cubital cell, see Text-figs. 12, 14, 18. Cells between transverse and apical veins are apical cells. When desirable these are numbered starting with the one nearest to the fore margin of the wing. When the terms "fore (anterior) margin" and "hind (posterior) margin" of wings are used in the present paper, they refer to directions parallel with the longitudinal axis of the body, the wings being spread out in flying position. The "fore margin" is also called the "costal margin" in spite of the absence of a true costa. In the fore wing there is an almost exactly straight suture extending from the wing base to a point on the hind margin of the wing. This suture, the "corioclaval suture" or "claval suture", divides the fore wing into two parts, a larger anterior part, corium, and a posterior part, clavus. In corium there are usually a number of transverse veins. By a transverse row of such veins the corium is often divided into a larger, proximal, more leathery part, and a distal, more membranous, so-called apical part, or membrane. The veins situated in clavus (anales) are also called the

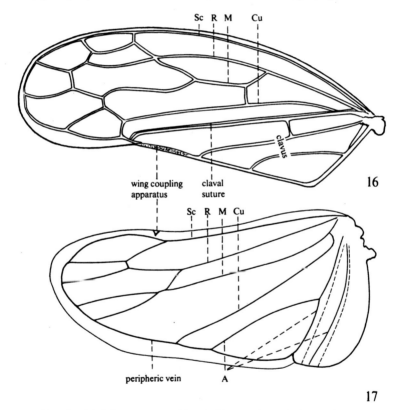

Text-fig. 16. *Scleroracus russeolus* (Fallén) (Deltocephalinae), fore wing.
Text-fig. 17. Same, hind wing.

claval veins. The free border of clavus is usually divided by an angular corner into two parts: a proximal part along the scutellum, the scutellar border, and a distal part, at rest contiguous with the corresponding border of the other fore wing, the commissural border. In Typhlocybinae and some other Cicadellidae, a wax gland is situated somewhat proximally of the middle of the costal border of the fore wing. Its secretion covers an oval part of the wing surface, the wax area (Text-fig. 18).

In the hind wing the apices of the longitudinal veins may or may not reach the apical margin of the wing. If not, they may end in a "peripheric vein" running approximately parallel with the hind and apical margins (Text-figs. 15, 17, 19). Normal wings have a wing-coupling apparatus for holding the fore and hind wing together during flight. It consists of a longitudinal fold or a quite short backward-directed lobe on the dorsoanterior border of the hind wing, and a longitudinal fold on the ventral commissural border of the fore wing. In Fulgoromorpha the structure of this apparatus is much more complex than in the Cicadomorpha. Additional wing-coupling mechanisms are found in many forms, such as curved spines or hooks along the anterior border of the hind wing. In Cercopidae these hooks are situated on a dilated part of the anterior

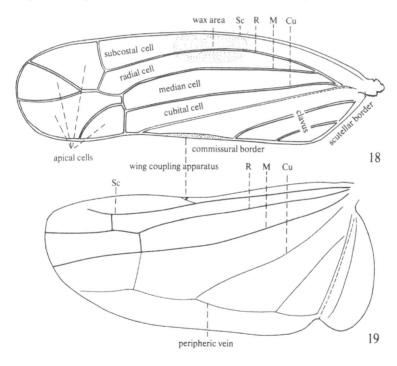

Text-fig. 18. *Eupteryx atropunctata* (Goeze) (Typhlocybinae), fore wing.
Text-fig. 19. Same, hind wing.

19

border of the hind wing proximally of the middle. In b rachypterous forms the claval suture, the veins and the wing-coupling apparatus may be distinct or more or less reduced depending on the degree of reduction of the wings.

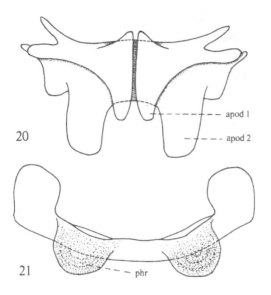

20

21

Text-fig. 20. *Eupelix cuspidata* (Fabricius) (Dorycephalinae), first two abdominal sterna with apodemes (apod 1, apod 2) from above.
Text-fig. 21. *Kybos butleri* (Edwards) (Typhlocybinae), 3rd abdominal tergum with phragma lobes (phr) from below.

Abdomen (Text-figs. 20, 21)

The abdomen is longish, cylindrical, conical, or with a triangular transverse section in most Auchenorrhyncha. In all males so far examined, and perhaps also most females, the first abdominal segment or segments I and II together contains a sound-producing apparatus, the so-called tymbal organ. Its essential parts are the tymbals, a pair of convex plates, not always distinctly demarcated, on the sides of the tergum of the first abdominal segment, and the tymbal muscles, a pair of dorso-ventral muscles belonging to the same segment. By the action of these muscles the tymbals are set into vibration producing sound. Additional muscles in abdominal segments I and II probably serve to change the curvature and tension of the tymbals, thus altering the frequency of the sound emitted. Endoskeletal processes, apodemes, of 1st and 2nd abdominal sterna, and similar processes from the tergal phragmas of the 2nd and 3rd abdominal segments, serve as attachments for these muscles. These processes are much varying in shape and size, and in the males are useful as species characters in some groups, e. g. *Macrostelis*, Idiocerinae and Typhlocybinae (Text-figs. 20, 21).

A functional tymbal organ seems to be absent in females of Cicadidae, and in other groups the structure of the organ is usually much simpler in females than in males.

A tympanal organ is present in the second abdominal segment of both sexes in the Cicadidae. A chordotonal organ has been found in the corresponding place in *Ribautiana ulmi* (L.) (Typhlocybinae) (Vondráček, 1949).

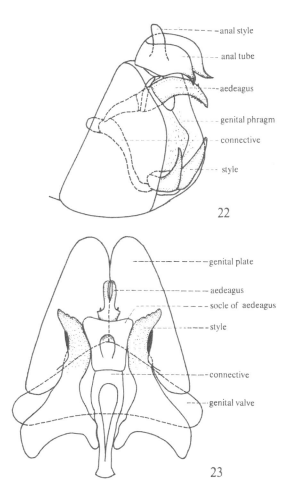

Text-fig. 22. *Stiroma bicarinata* (Herrich-Schäffer) (Delphacidae), pygofer of male obliquely from the left and behind.
Text-fig. 23. *Jassargus flori* (Fieber) (Deltocephalinae), sternal parts of male genital capsule as seen from above, tergal parts removed.

Male genitalia (Text-figs. 22–25)

In the males the 9th abdominal segment, the genital capsule or genital segment, contains the copulatory organs. These are partly strongly sclerotized and their structure is usually characteristic for and little varying within the species. The most important parts are the following: 1. Aedeagus. This is the tubular sclerotized continuation of the ductus ejaculatorius. It is sometimes partly or entirely enclosed in a penis-sheath, or

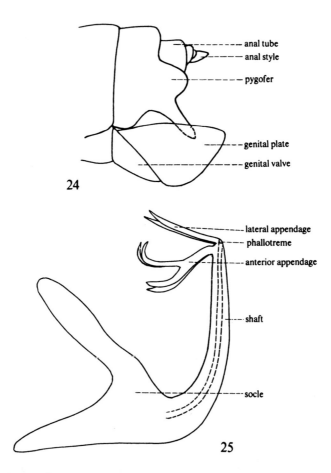

Text-fig. 24. *Graphocraerus ventralis* (Fallén) (Deltocephalinae), apex of male abdomen from the left.
Text-fig. 25. *Edwardsiana ishidae* (Matsumura) (Typhlocybinae), aedeagus from the left.

theca, a fold or sheath arising from the so-called phallobase. Both aedeagus and theca can either be simple or armed with appendages or processes varying much in shape. The terminal opening of the aedeagus is the phallotreme. 2. Parameres, or styles, a pair of appendages on inside of the genital plates or, in Delphacidae, behind the genital phragm. They are attached to the aedeagus by the connective, an often T- or Y-shaped, or trapezoidal, system of longitudinal and transverse bars. In Cicadomorpha (except Cicadidae) the parameres each consist of two parts, a basal genital plate, often covering and concealing the remaining genitalia from below with the one of the other side, and a

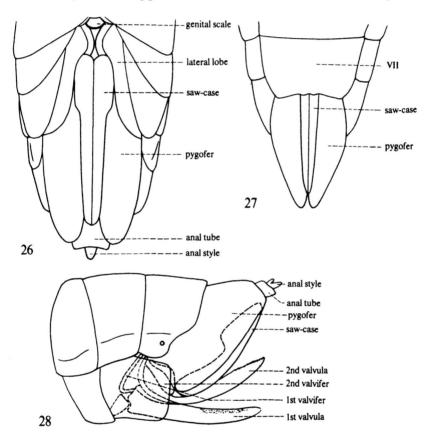

Text-fig. 26. *Delphax crassicornis* (Panzer) (Delphacidae), posterior part of female abdomen from below.

Text-fig. 27. *Athysanus argentarius* Metcalf (Deltocephalinae), apical part of female abdomen from below.

Text-fig. 28. Same, apical part of female abdomen from the left.

style. Thus the latter is only partly homologous with the styles of Delphacidae. In the Delphacidae and others the tergal, pleural and sternal parts of the genital segment are fused into a closed ring, the pygofer. In others the 9th abdominal sternum is a well-defined unpaired plate, the so-called genital valve, lying in front of the genital plates. The lateral parts of the genital segment, the pygofer lobes or lateral lobes, are often armed with processes characteristic in shape. In Delphacidae a transverse wall, the genital phragm, divides the room inside the genital capsule into two chambers, an anterior one enclosing the aedeagus at repose, and a posterior very shallow chamber opening caudally, in which the styles are situated (Text-fig. 22). The styles are basally connected with the connective through an opening in the lower part of the genital phragm, while the aedeagus can be extended by another opening in its upper part (as shown in the figure). The 10th abdominal segment is the anal tube which carries the terminal 11th segment, the anal style. A transverse intersegmental plate existing between the 9th and 10 th segments in some groups is called the anal collar. Like the anal tube, the anal collar may be armed with sclerotized hooks or other processes having a function during copulation and sometimes offering characters useful for separating species. In the Cicadidae, the styles, genital plates, connective and genital valve are all absent.

Female genitalia (Text-figs. 26–28)

The tergum of the 9th abdominal segment forms the pygofer. Caudad to the pygofer the 10th and 11th abdominal segments are represented by the anal tube and the anal style, respectively. Ventrally, in a groove of the pygofer, the ovipositor is situated. It consists of three pairs of oblong valves or valvulae, one pair emerging from the 8th and two pairs from the 9th abdominal segment. Basally they are attached to the so-called valvifers, paired rests of the 8th and 9th abdominal sterna. In the Delphacidae the valvifers of the 8th abdominal sternum are visible as two lobes flanking the proximal part of the ovipositor. These are called the lateral lobes (Text-fig. 26). In the same family an unpaired sclerite belonging to the 7th sternum is usually present in front of the base of the ovipositor. This "genital scale" is not always visible without dissection. The posterior pair of valves of the 9th sternum encloses the other two pairs of valves, the "saw", and is often termed the "saw-case". The valves composing the saw are generally more or less sabre-shaped. The outer pair belonging to the 8th sternum is called the anterior valves or first valvulae, the inner pair belonging to the 9th sternum is the median or inner valves or second valvulae. The latter are generally fused one to another, at least basally. In some families the ovipositor is absent or short and broad, tuberculiform. In Cicadellidae the hind margin of the 7th abdominal sternum is often incised or notched, or armed with caudal processes offering characters of diagnostic value.

24

Other morphological features

Many Auchenorrhyncha are richly equipped with wax glands. Thus the caudal end of *Cixius* females is usually enclosed in a white envelope of wax. Sensory hairs of various kinds are frequent on different parts of the body. In Fulgoromorpha, especially in the larvae, sensory hairs are located in special sensory pits.

Larval taxonomy

With certain exceptions, morphology and taxonomy of larval Auchenorrhyncha have been little studied so far. Vilbaste (1968) constructed keys to families of the larvae of Fulgoromorpha and also a key to North European genera of Delphacidae. The latter was largely based on the number and arrangement of sensory pits on the body segments. The key to families is reproduced here. We refrain from using his key to delphacid genera as several Fennoscandian species of Delphacidae were not considered by Vilbaste. Linnavuori (1951) separated the larvae of *Cixidia confinis* (Zett.) and *lapponica* (Zett.). Wagner (1950) keyed the larvae of the salicicolous *Macropsis* species of North and Central Europe on the basis of the shape of abdominal terga and body pilosity. Ossiannilsson (1950b) keyed the six species of Cercopidae living in the vicinity of Uppsala. His key used the length of rostrum, the shape of postclypeus and pigmentation as separating characters. Many descriptions of larval instars of various species of Auchenorrhyncha have been published. However, as many others are still unknown complete keys cannot be made at present.

Bionomics of the Auchenorrhyncha

Most Auchenorrhyncha, larvae as well as adults, live on the sap or cell content of living plants which they suck up by aid of their maxillary stylets. Many species are specialized on a certain plant genus or even on one single plant species. Others, such as our common meadow spittlebug, *Philaenus spumarius,* are less particular, a complete list of their food-plants being fairly long. We use the term "food-plant" for any plant species which can temporarily serve as a nutrition source for larvae or adults of a certain insect. Plants, on which an insect species can reproduce during at least one generation, are "host-plants" of that species. Many leafhoppers suck out the cells in the leaves of plants, and on the points where they have been feeding whitish spots appear, consisting of air-filled parenchyma cells. The leaves of roses and elms are often covered with such spots caused by the feeding of *Edwardsiana rosae* and *Ribautiana ulmi*. Other species penetrate the xylem or the phloem of plants with their stylets. Most vectors of plant viruses are found among the "phloem feeders". Some species belonging to the family

25

Achilidae live in crevices of dry or decaying wood, under the bark of dead trees, etc., and these seem to live on the sap of dead plant parts or possibly on the mycelium of fungi (Polyporaceae). Cicadas, cixiids, and certain Cercopidae are subterraneous during larval development, while the adults live in the open.

In the bodies of most Auchenorrhyncha – except Typhlocybinae – symbiontic yeasts have been found. These are usually localized in special organs, so-called mycetomes. The symbionts are "hereditary", i. e. transmitted from parents to offspring with the eggs. Their role in the metabolism of their hosts is still unknown.

The eggs of the species of the family Tettigometridae – not represented in Fennoscandia and Denmark – are deposited freely. In this family the ovipositor is reduced. The Cixiidae place their eggs in the soil by aid of their graver-like ovipositor. The Issidae (and Dictyopharidae) use their strongly modified ovipositor for making a mantle of soil particles and a secretion for the eggs, which are deposited on the ground (Müller, 1942). The females of our Delphacidae and Cicadomorpha deposit their eggs single or in groups or rows into branches, twigs or leaves of their host-plants. The mechanism of hatching of the eggs has been described in detail by Müller (1951). The young larvae of Cicadidae fall to the ground and commence their subterranean life, but the larvae of other Auchenorrhyncha continue living on the plants, where oviposition has taken place. Most species are univoltine in our climatic conditions, but others manage to produce 2 or 3 generations per annum. On the other hand a development extended over two or several years is known to occur in the Cicadidae. Probably also *Ledra aurita* has a biennial life-cycle. Most Delphacidae and some other species hibernate in the larval stages. The majority of the remaining Auchenorrhyncha hibernate in the egg stage, but several species pass the winter as adults, often on conifers or other evergreen plants. The number of larval instars is reported to be six in Cicadidae, but in most other Auchenorrhyncha so far studied five preadult instars have been found, the egg stage not included. The post-embryonic development is gradual (paurometabolism).

Most Auchenorrhyncha are active at day. During day-time the adults generally prefer staying on the most sun-exposed parts of their food-plants. Exceptions certainly exist. Certain foreign species, as *Neoaliturus tenellus* (Baker), *Macrosteles fascifrons* (Stål) and *Empoasca fabae* (Harris), may migrate very long distances, if local conditions are unsuitable. Such migrations usually take place at night. Experiments with *Zygina hyperici* (H.-S.) and *Streptanus marginatus* (Kbm.) show that the males of these species sing during the night as well as during the day, irrespective of light conditions, but the temperature optimum for this kind of activity is fairly high (Ossiannilsson, 1949a). Our species are usually vivacious but not especially shy insects. Their capability of leaping makes them able to escape from a threatening danger very rapidly, but usually they do not use this capacity until the danger is imminent. First they will try to hide on the underside of a leaf or behind the stalk, where they have been sitting. Many species of e. g. Typhlocybinae and Idiocerinae fly readily and use their leaping ability only for taking wing.

The larvae usually stay out of the way on the underside of leaves etc. The larvae of froghoppers (Cercopidae) are remarkable by their peculiar manner of living. The conspicuous froth-lumps called "cuckoo spit" on various plants are a product of these larvae. The ragged robin, *Lychnis Flos-cuculi* L. (Swedish: gökblomster) has got its Latin and Swedish names as an allusion on the supposed association of this frothy substance with cuckoos, in popular belief, apparently because this herb is one of the preferred host-plants of *Philaenus spumarius*. On the other hand frogs have also been suspected to be producers of this foam, which explains the British name "froghoppers" and the Swedish "grodspott". The cercopid larvae can be found in these lumps sitting with their heads downwards and with their stylets stuck into the vessels of the host-plant. The liquid is their excrement fluid. Since Šulc (1911) this fluid has generally been supposed to be a sort of soap solution. According to Ziegler & Ziegler (1958) this is wrong. The essential properties of the fluid result from certain albuminous substances. Air from the respiratory organs is continuously pumped into the fluid. The froth thus produced runs down over the insect covering it on all sides. It protects its inhabitant against drought and against a number of enemies. However certain predaceous wasps and bugs and some birds do not hesitate to attack the froghoppers in their hiding-place.

The sound-production of the Auchenorrhyncha has been briefly mentioned above in connection with the structure of the tymbal apparatus. It appears from studies so far published that each species can produce several different calls apparently with different biological significance (Ossiannilsson, 1949a, 1953a; Strübing, 1958a and b, 1959, 1965; Claridge & Howse, 1968). From the work of Strübing we know that the capacity of sound-production is common also in females. The calls of the latter usually differ considerably from those of males. Species identification by aid of calls seems to be possible in many cases (see Strübing, 1970; Claridge & Reynolds, 1973). As the calls of our small species of Auchenorrhyncha are very weak, communication between individuals by these calls is probably brought about through the solid substrate, not by the air (Strübing, 1977).

Parasites

The most important parasites of our Auchenorrhyncha belong to three different insect groups: (1) flies of the family Pipunculidae, (2) wasps of the family Dryinidae (*Gonatopus, Dicondylus* and other genera), and (3) Strepsiptera (in Europe the species *Elenchus tenuicornis* Kirby). Pipunculid larvae live singly in the body of the leafhopper, particularly in the abdomen. The abdomen of a "pipunculized" individual usually expands somewhat and a tendency to develop a reddish pigment in the integument is often observed. Leafhoppers and other Auchenorrhyncha attacked by Dryinidae are easily recognized by carrying an external gall-like cyst (thylacium) on the outside of their body. This cyst may be as large as the abdomen of the host. It is usually black or yellow. One leafhopper may carry more than one cyst. In a "stylopized" specimen, i. e. an individual attacked by *Elenchus,* a small button-shaped protuberance can often be observed somewhere on the body. This is the anterior part of the so-called puparium of the parasite, the major part of which is situated inside the body of the host. The parasitism of *Elenchus* on *Chloriona* was studied by Lindberg (1939). In Finland only Delphacidae are attacked by *Elenchus* (Pekkarinen & Raatikainen, 1973). – A common result of the attack of these different parasites, if the host reaches the adult stage, is "parasitic castration", modification of sexual characters, reduced genitalia and appearance as "intersexes". The reduction of the copulatory organs may be more or less radical. Several "new species" have been described from specimens "castrated".

Economic importance of Auchenorrhyncha

Many Auchenorrhyncha are pests of cultivated plants. Direct damage is caused by the insect feeding, resulting in loss of sap which can be disastrous in periods of drought. Examples of this kind of damage are *Empoasca facialis* Jacobi, a pest of cotton, peanuts and other cultivated plants in Central and South Africa, and the sugar-cane planthopper, *Perkinsiella saccharicida* Kirk., a dangerous pest of sugar-cane in Java, Formosa and especially Hawaii. *Empoasca fabae* (Harris), *E. facialis,* and *E. lybica* de Bergevin and other species attacking various cultivated plants in many parts of the world cause "hopperburn", blockage of vascular tissues by reaction products of their saliva. This results in discoloration of leaves, deformation and shortening of shoots, checking of fruiting, stunting and premature death of plants. Hopperburn and damage by loss of sap are caused by "phloem-feeders" and "xylem-feeders". "Mesophyll feeders" (most Typhlocybinae) emptying parenchyma cells one by one reduce the assimilative surface of the leaves. Losses by this kind of attack are probably not serious, but the leaves of ornamental plants like roses may deteriorate considerably in appearance, if the leafhoppers are not controlled. Another type of damage is caused by the oviposition of Cicadidae, Membracidae, Cercopidae and certain other Auchenorrhyncha. The ovipositor of the females makes wounds, which may cause withering and breaking of branches. Our common *Cicadella viridis* (L.) makes damage of this kind on fruit trees and other plants in many parts of the world, but so far not here.

Serious direct damage on cultivated plants caused by Auchenorrhyncha is probably rare in the Scandinavian countries. In summers with persistent dry weather the "dwarf leafhopper", *Macrosteles laevis* (Ribaut), has been reported as a pest on cereals and other grasses in Sweden (Tullgren, 1925). The larvae of our common meadow spittlebug, *Philaenus spumarius* (L.), cause deformations of shoots on many of its numerous host-plants. This may be a nuisance in gardens with e. g. dahlias and pinks *(Dianthus).* The same species has been reported attacking cultivated strawberries and raspberries in our countries. In North America, where the species is not indigenous but introduced, the "European spittle insect" is a serious pest on alfalfa, red clover and *Melilotus*. Here it is not considered as a problem in these crops so far. Mesophyll feeders as *Empoasca vitis* (Göthe), *E. solani* (Curtis) and *Eupteryx atropunctata* (Goeze) on beets and potatoes, *Edwardsiana rosae* (L.) on *Rosa, Rubus, Fragaria,* and *Ribautiana ulmi* (L.) on *Ulmus,* sometimes change the colour of the leaves of their host-plants conspicuously and may reduce photosynthesis.

The greatest economic importance of Auchenorrhyncha is as vectors of viral and mycoplasmal plant diseases. Plant viruses transmitted by leafhoppers, froghoppers are all "circulative", which means that the virus imbibed by the insects with plant sap penetrates the gut wall, follows the blood and enters the salivary glands. A latent period in the vectors must be passed before they become infective. Then the virus is ejected into plants with the saliva. Some circulative viruses are also "propagative", in other words they multiply in the bodies of their vectors. Passage of virus from one generation to the next through the egg has been demonstrated in a number of leafhoppers and planthoppers. Usually plant viruses transmitted by auchenorrhynchous Homoptera are also vector-specific, being dependent on a single or a few closely related insect species for spread from plant to plant. The beet leafhopper, *Neoaliturus tenellus* (Baker) transmits the Californian curly top virus of beets and other crops. Aster yellows, a disease of several different plant species occurring in a few different strains, was believed to be caused by virus, but is now supposed to be associated with unicellular organisms called mycoplasma. The various strains of aster yellows are circulative and propagative and their vectors are species of *Macrosteles*. Heinze (1959) listed 128 species of Auchenorrhyncha as vectors of plant virus but many of the diseases in question are now reckoned to be caused by mycoplasma. Practically all of these vectors are phloem feeders.

In Sweden and Finland, plant diseases transmitted by planthoppers or leafhoppers are being studied by Lindsten and Raatikainen and their co-workers. Most important diseases so far studied affect cereals and are carried by planthoppers (Delphacidae). More detailed information is given under the various vectors.

Key to families of Auchenorrhyncha (adults)

1   Median coxae long, shaped more or less like fore coxae, but with bases widely separated (in North-European families). A pair of tegulae present immediately in front of bases of fore wing (Text-fig. 10), (sometimes the tegulae are hidden under posterior margin of pronotum). Clypeus with upper margin below level of lower margin of compound eyes. Antennae below compound eyes, or below level of their lower margin. Two anal veins in clavus apically united (Text-figs. 10, 12)
(Infraorder **Fulgoromorpha**) 2

–   Median coxae short, not shaped as fore coxae, their bases close together. Tegulae absent. Clypeus (frontoclypeus) with upper margin above level of lower margin of

29

compound eyes. Antennae situated between compound eyes and clypeus (frontoclypeus). All veins in clavus free (Text-figs. 14, 16, 18).

(Infraorder **Cicadomorpha**) 5

2 (1) Hind tarsi with second segment not especially small, apically with a row of small spines 3

- Hind tarsi with second segment very small, apically without or at most with 2 spines. Head with eyes almost as broad as pronotum. Hind margin of pronotum almost straight. Scutellum short, triangular **Issidae**

3 (2) Anal vein reaching apex of clavus. Fore wings distally of apex of clavus widening, in resting position apical part of one of them overlapping that of the other side **Achilidae**

- Anal vein ending in commissural border proximally of claval apex 4

4 (3) Hind tibia with a large movable spur (post-tibial calcar) (Text-fig. 7) **Delphacidae**

- Hind tibia without a movable spur **Cixiidae**

5 (1) Pronotum with a posterior process projecting backwards over mesonotum and abdomen. Vertex and anterior part of pronotum vertical. Face horizontal, directed to the substratum **Membracidae**

- Pronotum without a posterior process, its anterior part not vertical 6

6 (5) Three ocelli present, arranged in a triangle on the vertex. Anterior femora swollen, armed with spines beneath **Cicadidae**

- Two ocelli, or ocelli absent. Anterior femora not conspicuously swollen 7

7 (6) Hind tibiae without keels, laterally with one or several stout fixed spines, but without mobile setae (Text-fig. 8) **Cercopidae**

- Hind tibiae with one or more longitudinal keels bearing rows of mobile setae (few and poorly developed in *Ulopa*) (Text-fig. 9) **Cicadellidae**

Key to families of Auchenorrhyncha (larvae)

Partly after Vilbaste (1968)

1 Postclypeus distinct from frons. Median coxae long, their bases widely separated. Body usually with sensory pits (Infraorder **Fulgoromorpha**) 2

- Postclypeus not distinct from frons. Median coxae short, their bases set close together. Body without sensory pits (Infraorder **Cicadomorpha**) 5

2 (1) Hind tibiae with foliaceous or awl-like spur ("post-tibial calcar") at apex (in young instars usually very small) **Delphacidae**

- No post-tibial calcar present 3

3 (2) Hind tarsi with second segment with teeth only on sides **Issidae**

- Hind tarsi with second segment with a row of teeth at apex 4

4 (3) Abdominal terga VI–VIII with "wax spots" (as white patches on a brownish ground). The rows of sensory pits on abdominal terga IV–V extend almost to the mid-line. Edges of vertex rounded, not bordered by keels. Ground colour brown. Subterraneous **Cixiidae**
   – Abdominal terga VI–VIII without "wax spots"; white spots, if present, situated on the sides of the terga. Sensory pits on abdominal terga IV–V situated laterally. Vertex bordered with keels. Ground colour whitish or greyish. In crevices in dead wood **Achilidae**
5 (1) Fore legs modified into a strong digging-tool, similar to the front pincers of a crab **Cicadidae**
   – Fore legs not modified for digging 6
6 (5) Pronotum strongly elevated, gibbous. Last abdominal segment slender, conical, approximately as long as rest of abdomen **Membracidae**
   – Pronotum not strongly elevated. Last abdominal segment not or little longer than the preceding segment 7
7 (6) Terga and pleura of abdominal segments III–IX curved beneath the abdomen as membranous extensions, concealing the true sternum. In "cuckoo-spit" **Cercopidae**
   – Terga and pleura of abdomen not curved beneath the latter **Cicadellidae**

Family Cixiidae

Wings large, broad, usually transparent. Frons broad, with 3 carinae, clypeus rather small. As a rule with 3 ocelli. Antennae usually small, first segment short, second segment larger, globular. Pronotum usually broad and short, mesonotum large, with 3 or 5 carinae. Tegulae present. Legs thin, hind tibiae with or without lateral spines. Aedeagus enclosed in a theca. Ovipositor of female graver-like, protruding from apex of abdomen (Text-fig. 40).

Larvae living on roots in the soil, adults on trees and shrubs. In Denmark and Fennoscandia 3 genera.

Key to genera of Cixiidae

1 Scutellum with five longitudinal carinae   *Pentastiridius* Kirschbaum (p. 42)
   – Scutellum with three longitudinal carinae   2
2 (1) Fore wings with tubercles on apical margin between veins
                                                   Tachycixius W. Wagner (p. 41)
   – Fore wings without tubercles on apical margin between veins
                                                   Cixius Latreille (p. 32)

31

Genus *Cixius* Latreille, 1804

Cixius Latreille, 1804: 310.

Type-species: *Cicada nervosa* Linné, 1758, by subsequent designation.

Head a little narrower than pronotum, its hind margin arched, concave. Frons and clypeus with a common median carina. Scutellum with 3 carinae. In our species, veins of fore wings with conspicuous dark setiferous tubercles. Dark spots along costal margin distinctly larger than tubercles on veins. Fore wing in first apical cell with a thickening, the pterostigma or stigma.

Key to species of *Cixius*

1 Transverse keel between frons and vertex almost obsolete; frons and vertex in lateral aspect broadly rounded into each other. Anterior keel of vertex rounded, somewhat flattened   3. *distinguendus* Kirschbaum
- Transverse keel between frons and vertex more or less distinct. In lateral aspect, frons and vertex form a more or less distinct angle. Anterior keel of vertex sharp 2
2 (1) Clypeus yellowish. Large species, 5.8–8.0 mm   3
- Clypeus and frons black with light carinae   4
3 (2) Fore wing broad and comparatively short, index length: maximal width about 2.7. Hind wing with a distinct fuscous streak along apical part of costal margin   1. *cunicularius* (Linné)
- Fore wing narrower, index length: maximal width as a rule about 3.0. Hind wing membrane without a fuscous streak along apical part of costal margin, colourless, or with fuscous areas inside apical cells not touching veins   2. *nervosus* (Linné)
4 (2) Fore wing membrane fuscous   4. *similis* Kirschbaum, variety
- Fore wing membrane ± colourless, with or without dark markings   5
5 (4) Lateral spines of aedeagal sheath (theca) each with 2 or more apical points (Text-figs. 41, 42)   4. *similis* Kirschbaum
- Lateral spines of theca simple at apex   6
6 (5) Anal tube apically with a pair of semilunar, laminar projections visible from above (Text-figs. 52, 53)   6. *cambricus* China
- Apical projections of anal tube inconspicuous, not visible from above (Text-figs. 47, 48)   5. *stigmaticus* (Germar)

1. *Cixius cunicularius* (Linné, 1767)
Plate-fig. 1, text-figs. 29–32.

Cicada cunicularia Linné, 1767: 711.

Carinae of head sharp. Black. Frons with yellowish carinae, clypeus yellowish. Pronotum, tegulae and legs partly dirty yellow. Pigmentation of fore wing varying. In the typical form the fore wings are colourless with the base, a more or less distinct transverse series of spots proximally of middle, and a broad transverse band from stigma to apex of clavus, dark brown. In f. *fusca* Fieber the entire fore wing proximally of the last-mentioned transverse band is dark brown. Hind wing apically with an oblong dark patch along its costal margin. Aedeagus as in text-figs. 29,30, style as in text-fig. 31, anal tube as in text-fig. 32. Overall length 6–8 mm.

Distribution. Fairly common in Denmark (SJ, EJ, WJ, NEJ, F, LFM, NWZ, NEZ,

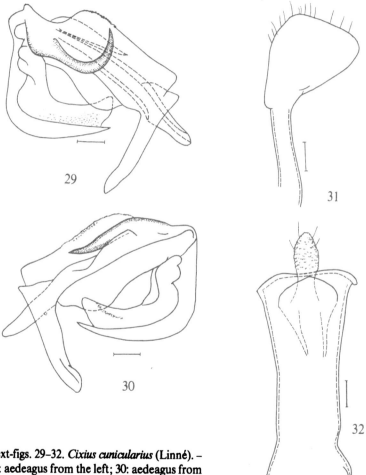

Text-figs. 29–32. *Cixius cunicularius* (Linné). – 29: aedeagus from the left; 30: aedeagus from the right; 31: right genital style; 32: male anal tube. Scale: 0.1 mm.

B). – Common in Sweden, found in all provinces. – Also known from most districts in Norway. – Common in East Fennoscandia, recorded from all provinces except ObS and LkE. – Widespread in Europe, also in Algeria, Georgia, Kazakhstan, Manchuria, and m. Siberia.

Biology. Adults on foliiferous trees and bushes, July–September.

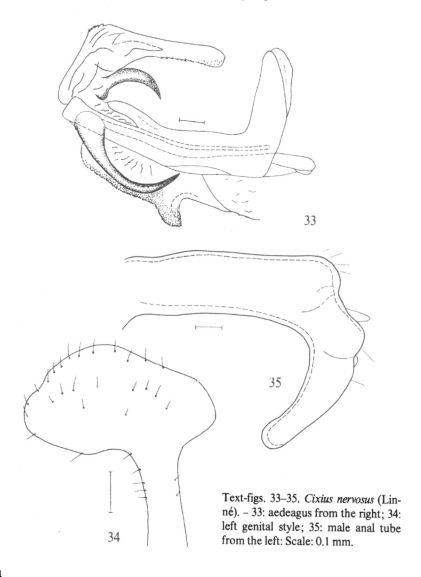

Text-figs. 33–35. *Cixius nervosus* (Linné). – 33: aedeagus from the right; 34: left genital style; 35: male anal tube from the left: Scale: 0.1 mm.

2. Cixius nervosus (Linné, 1758)
Plate-fig. 25, text-figs. 33–35.

Cicada nervosa Linné, 1758: 437.

Carinae on head sharp. Black. Frons black, its carinae and clypeus orange or yellowish. Legs partly dirty yellow. On the fore wings is the very base largely fuscous. Normally a dark transverse band is present basad of middle of the fore wing. More apically there are some irregular dark spots, especially around the transverse veins. Anal tube of male (text-fig. 35) apically with a pair of large appendages directed obliquely cephalad and ventrad. Style as in Text-fig. 34, aedeagus as in Text-fig. 33. Our largest species. Length 6–8.5 mm.

Distribution. Common in Denmark. – Fairly common in Sweden, found in most provinces up to P. Lpm. – Norway: HEn, VAy, Ry, HOy, SFi, MRy. – Rare in East Fennoscandia (Al, Ab, N; Vib, Kr). – Widespread in Europe, also in N Africa, W Asia, Japan.

Biology. Adults on foliiferous trees and bushes, June–September. In Central Europe hibernation takes place in larval stages (Müller, 1957).

3. Cixius distinguendus Kirschbaum, 1868
Text-figs. 36–40.

Cixius distinguendus Kirschbaum, 1868: 48.
Cixius intermedius Scott, 1870: 147.
Cixius brachycranus Scott, 1870: 148.

Transverse keel between frons and vertex almost obsolete. In lateral aspect, frons and vertex evenly rounded into each other. By this character *C. distinguendus* differs from all other European species of *Cixius*. Clypeus yellowish to brownish, frons dark brown to fuscous, above lighter, carinae yellowish brown. Markings of fore wings as in *nervosus* but usually less well-marked. Aedeagus as in Text-figs. 36 and 37, styles as in Text-fig. 38, anal tube of male as in Text-fig. 39, apex of female abdomen as in Text-fig. 40. Overall length 6–7.7 mm.

Distribution. Rare in Denmark, only found in B: Jons Kapel 11.IX.1974 (L. Trolle). – Not uncommon in Sweden, Sk. – Hls. – In Norway found in Ry: Ryfylke 1912 (specimen in Stavanger Museum, collector anonymous), SFi: Aurland, Vassbygda (Knaben), Luster 23.VII.1945 (Knaben); VAy: Valand 1943/44 (Holgersen); HOy: Bruvik, Eikemo, Eik 3.VIII.1968 (L. Greve). – East Fennoscandia: rare, found in Al, Ab, N, Ta, Sa, Vib, and Kr. – Widespread in Europe, also in Kazakhstan.

Biology. In leafy woods, also in the herbaceous vegetation under the trees, adults in July–September.

35

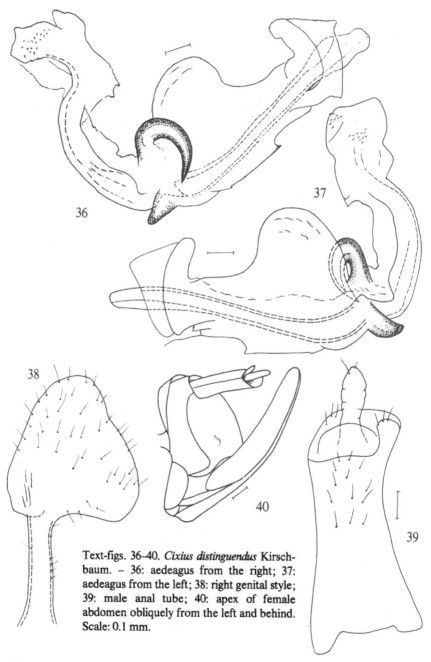

Text-figs. 36–40. *Cixius distinguendus* Kirschbaum. – 36: aedeagus from the right; 37:
aedeagus from the left; 38: right genital style;
39: male anal tube; 40: apex of female
abdomen obliquely from the left and behind.
Scale: 0.1 mm.

4. *Cixius similis* Kirschbaum, 1868
Text-figs. 41–44.

Cixius similis Kirschbaum, 1868: 48–49.
Cixius distinguendus J. Sahlberg, 1871: 382 (nec Kirschbaum).

Carinae of head sharp. Black. Frons and clypeus black with reddish-yellow carinae. Pronotum, tegulae, legs and segment margins of abdomen in their major parts reddish-yellow. Fore wings hyaline, colourless with diffuse smoke-coloured spots, or practically entirely smoke-brownish with a light spot at the stigma. Aedeagus as in Text-figs. 41 and 42, style as in Text-fig. 43, anal tube of male as in Text-fig. 44. Overall length 5–6.5 mm.

Distribution. Fairly common in Denmark, found in most districts. – Common in Sweden, Sk. – T. Lpm. – Norway: found in HEn, On, VAi, Ry, HOy, STi, TRi, and Fø. – East Fennoscandia: common and found in almost all provinces. – Europe, Kazakhstan.

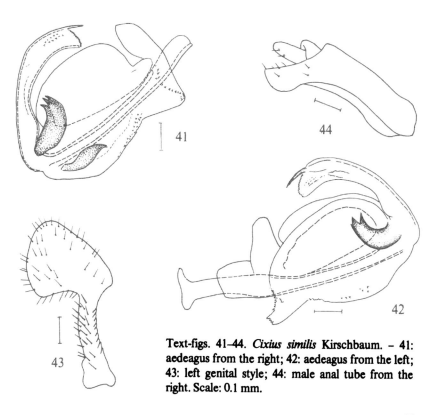

Text-figs. 41–44. *Cixius similis* Kirschbaum. – 41: aedeagus from the right; 42: aedeagus from the left; 43: left genital style; 44: male anal tube from the right. Scale: 0.1 mm.

Biology. On bogs and marshes among *Betula nana*, *Salix* spp., *Ledum*, *Myrica* etc, adults in May–August.

5. *Cixius stigmaticus* (Germar, 1818)
Text-figs. 45–48.

Flata stigmatica Germar, 1818: 199.

Carinae of head sharp. Frons and clypeus black with light carinae. Fore wings hyaline, colourless or very faintly yellowish, markings indistinct or absent. Stigma dark. Aedeagus as in Text-fig. 45, style as in Text-fig. 46, male anal tube as in Text-figs. 47 and 48. Overall length 6–7 mm.

Distribution. Rare in Denmark, only found in NWZ: Jyderup 21.VI.1915 (C. C. R. Larsen), and in SZ: Bregentved 22.VIII.1916 (O. Jacobsen). – Very rare in Sweden, one

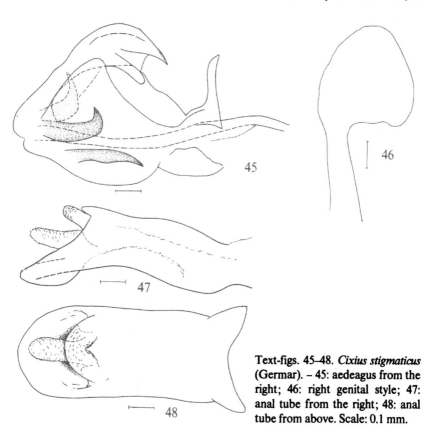

Text-figs. 45–48. *Cixius stigmaticus* (Germar). – 45: aedeagus from the right; 46: right genital style; 47: anal tube from the right; 48: anal tube from above. Scale: 0.1 mm.

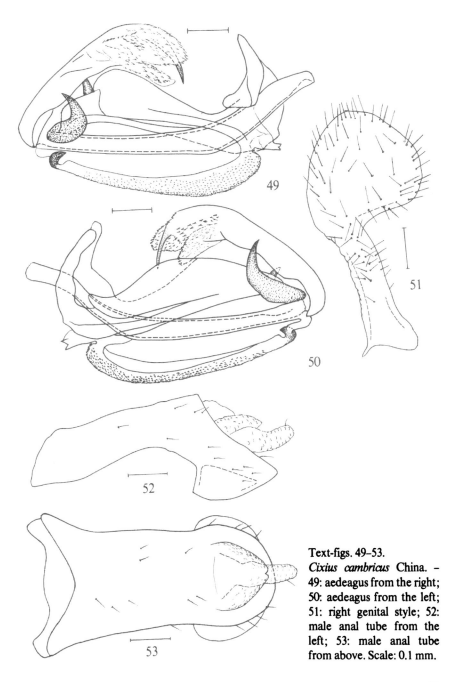

Text-figs. 49–53.
Cixius cambricus China. –
49: aedeagus from the right;
50: aedeagus from the left;
51: right genital style; 52:
male anal tube from the
left; 53: male anal tube
from above. Scale: 0.1 mm.

male only being found in Sk., Skäralid, 8.VI.1967 (K. Ander leg.). – Norway: two males were collected by students taking part in excursions in HOy: Tysnes, Ånuglo 4.VI.1966 and 29.V.1969. – So far not found in East Fennoscandia. – France, Central, Southern and Eastern Europe.

Biology. In shrubby localities, often on *Alnus* or *Salix* spp. (Dlabola, 1954).

6. *Cixius cambricus* China, 1935
Text-figs. 49–53.

Cixius cambricus China, 1935: 38.
Cixius borussicus W. Wagner, 1939: 107.
Cixius austriacus W. Wagner, 1939–107.

Carinae of head sharp. Frons and clypeus black, carinae light. Prothorax largely yellow, mesonotum black. Fore wings hyaline, colourless with or without a brownish

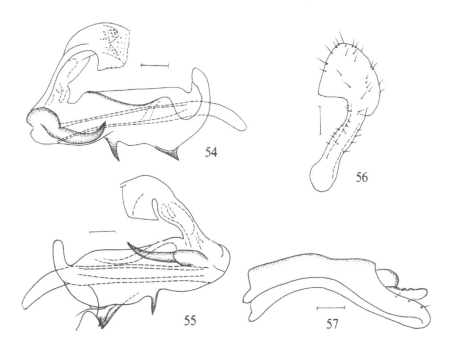

Text-figs. 54–57. *Tachycixius pilosus* (Olivier). – 54: aedeagus from the right; 55: aedeagus from the left; 56: left genital style; 57: male anal tube from the left. Scale: 0.1 mm.

40

transverse band at level of claval fork. Transverse veins in apical third of fore wing bordered with fuscous. Aedeagus as in Text-figs. 49 and 50, style as in Text-fig. 51, anal tube of male as in Text-figs. 52 and 53. Overall length 4.5–6 mm.

Distribution. Found in East Fennoscandia: Al, Ab, N, Kb (Huldén, 1975: 88). – So far not in Denmark, Sweden and Norway. – Widespread in Europe including Great Britain; also in Georgia and Azerbaijan.

Biology. A stenotope inhabitant of the "wood steppe" ("Waldsteppe"). Hibernation in larval instars (Schiemenz, 1969). The Finnish specimens were collected in July and August (Huldén, l. c.).

Genus *Tachycixius* W. Wagner, 1939

Tachycixius (ut subgenus) W. Wagner, 1939: 96.
Type-species: *Fulgora pilosa* Olivier, 1791, by subsequent designation.

As *Cixius*, but apical part of fore wing with tubercles. In our countries only one species.

7. *Tachycixius pilosus* (Olivier, 1791)
Text-figs. 54–57.

Fulgora pilosa Olivier, 1791: 575.
Flata contaminata Germar, 1818: 196.

Frons and clypeus brick reddish or yellowish brown. Vertex black between the orange-coloured carinae. Pronotum black with reddish yellow margins, mesonotum black with reddish yellow carinae. Fore wings hyaline, colourless with scattered orange-coloured spots, usually more dense or coalescent on the apical third of the wing. A more or less distinct transverse band dissolved in spots basad of middle. Base of fore wing generally dark. Dark spots along costal margin not larger than tubercles of veins. Setae on wing tubercles conspicuously long (if not rubbed off). Anal tube of male (Text-fig. 57) without apical teeth. Aedeagus as in Text-figs. 54 and 55, style as in Text-fig. 56. Overall length 4–6 mm.

Distribution. Fairly common in Denmark, especially in Jutland, found in SJ, EJ, WJ, F, LFM, SZ, and NEZ. – Sweden: rare, but locally abundant in Sk. and Bl., also found in Gtl. – Not found in Norway and Finland, nor in European Russia except Ukraine. Widespread in Central and Southern Europe, also found in Tunisia, Azerbaijan and the Nearctic.

Biology. Adults in May–July, especially on dry places and in localities with poor vegetation. Often on *Quercus* and *Betula* (Dlabola, 1954). Hibernation takes place in larval instars (Müller, 1957).

41

Genus *Pentastiridius* Kirschbaum, 1868

Pentastiridius Kirschbaum, 1868: 45.
  Type-species: *Flata pallens* Germar, 1821, by monotypy.

Head a little narrower than pronotum. Hind margin of head with an angular concavity. Frons and clypeus with a common median carina. Scutellum with 5 longitudinal carinae. Fore wing with a pterostigma as in *Cixius*. In Denmark and Fennoscandia 1 species.

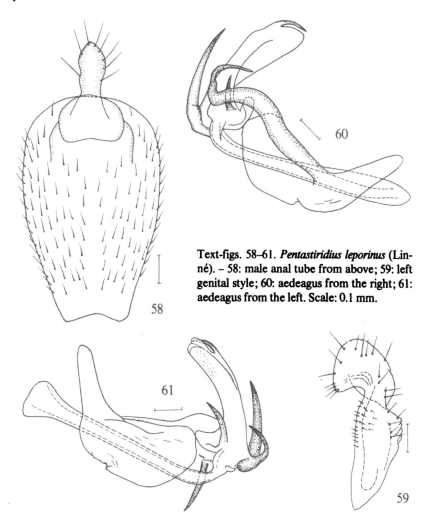

Text-figs. 58–61. *Pentastiridius leporinus* (Linné). – 58: male anal tube from above; 59: left genital style; 60: aedeagus from the right; 61: aedeagus from the left. Scale: 0.1 mm.

8. *Pentastiridius leporinus* (Linné, 1761)
Plate-fig. 2, text-figs. 58–61.

Cicada leporina Linné, 1761: 242.
Oliarus leporinus J. Sahlberg, 1871: 385.

Body black. Carinae and margins of head, margins of thorax, hind margins of abdominal segments, and partly also legs yellowish. Wings transparent, colourless, with veins and stigma more or less dark. Setiferous tubercles on fore wing veins small and indistinct. Male anal tube as in Text-fig. 58, style as in Text-fig. 59, aedeagus as in Text-figs. 60 and 61. Overall length 5.5–7.5 mm.

Distribution. Denmark: rare, only found in SJ: Nordby, Fanø 5.VIII.1911 (O. Jacobsen), and in LFM: Knuthenborg 26.V.1913 (A. C. Jensen-Haarup). – Rare also in Sweden: Sm., Gårdsby (J. A. Z. Brundin); Upl., Uppsala (Linné); Dlr., Sundborn, Karlsbyheden, Hedkarlsjön (A. Jansson), Ore, Djupaspen (T. Tjeder); Hls., Gnarp, Leska (Ossiannilsson); Nb., N. Luleå, S. Sunderbyn (H. Andersson). – So far not found in Norway. – Fairly rare in East Fennoscandia, found in Al, Ab, N, St, Kb, and Kr. – Europe, N. Africa, N. China, western parts of Asiatic USSR.

Biology. On *Phragmites* and other swamp grasses, adults in May–August. Prefers very wet biotopes (Wagner & Franz, 1961). Also in dry places (Linnavuori, 1952).

Family Delphacidae
Planthoppers

Most species of this family are small insects. Second segment of hind tarsi apically with a transverse row of small spines. Hind tibiae apically with a mobile spur, "post-tibial calcar" (Text-fig. 7). Claval veins not granulous, their common stem reaching commissural border proximally of claval apex (Text-fig. 12). Many species are wing-dimorphous. Frons laterally defined by a pair of lateral carinae. Most species have between these a single median carina which forks above in two branches continuing on vertex; in others there are two parallel median carinae instead of a single one. Sometimes the carinae are indistinct. Genital segment of males divided in two spaces by a transverse wall, the genital phragm (Text-fig. 22). Theca of aedeagus usually absent or rudimentary. A genital scale present in females of many species (Text-fig. 26). All of our species live on herbaceous plants. In Denmark and Fennoscandia 39 genera.

Key to genera of Delphacidae

1	Lateral carinae of pronotum reaching its hind border	2
–	Lateral carinae of pronotum not reaching posterior border, more or less curving outwards posteriorly	7

2 (1)   Frons broadest between compound eyes. Vertex extending only slightly in front of eyes, anteriorly rounded. First segment of hind tarsi with 4 + 2 apical spines. Genae often with a black spot (Text-figs. 3, 81)   3
–   Frons not broadest between compound eyes. First segment of hind tarsus with 5+ 2 or 6 + 2 apical spines   4
3 (2)   Theca of aedeagus absent or strongly transformed, not enclosing aedeagus
    Kelisia Fieber (p. 49)
–   Theca sheath-like, enclosing major part of aedeagus (Text-figs. 102, 109)
    Anakelisia W. Wagner (p. 58)
4 (2)   Frons at least two and a half times as long as broad, in lower half parallel-sided. Fore wings always longer than abdomen. Apex of fore wing angular
    Stenocranus Fieber (p. 62)
–   Frons shorter, broadest in lower half. Fore wings longer or shorter than abdomen; if they are longer, their apex rounded   5
5 (4)   Lateral carinae of pronotum less divergent, distance between their posterior ends subequal to length of median carina (Text-fig. 283). Vertex extending considerably in front of eyes   *Megamelus* Fieber (p. 103)
–   Lateral keels of pronotum more divergent, distance between their posterior ends considerably greater than length of median carina (Text-fig. 387). Vertex short, only extending a small distance in front of eyes   6
6 (5)   Styles and appendages of anal tube long (Text-fig. 365). Fore wings of brachypters dark with a few lighter spots   *Megamelodes* Le Quesne (p. 128)
–   Styles and appendages of anal tube short (Text-figs. 378, 388). Fore wings of brachypters unicolorous, without markings or with darker markings
    Delphacodes Fieber (p. 132)
7 (1)   Vertex distinctly narrowing in front, distinctly longer than broad (Text-fig. 236). Males (rarely females) macropterous, with pronotum and scutellum grey-green, females usually brachypterous, in life bright green   *Chloriona* Fieber (p. 95)
–   Vertex not distinctly narrowing in front, rarely distinctly longer than broad. Not green or grey-green   8
8 (7)   Basal segment of antenna about three times as long as its diameter   9
    –   Basal segment of antenna shorter   10
9 (8)   First antennal segment flattened, the second one cylindrical, shorter than first segment   *Delphax* Fabricius (p. 86)
–   First antennal segment conical, second segment cylindrical, longer than first segment   *Euides* Fieber (p. 93)
10 (8)   Median keel or keels of frons very indistinct, obsolescent or entirely absent   11
–   Median keel or keels distinct   13
11 (10)   Head as seen from above distinctly shorter than broad between eyes   12
–   Head about as long as broad between eyes, anteriorly with a well-defined black spot (Text-fig. 177)   *Stiromoides* Vilbaste (p. 77)
12 (11)   Median keel of frons almost entirely obliterated. Vertex, frons, and clypeus concolourous, brownish   *Eurysula* Vilbaste (p. 71)

-      Median keel of frons distinct in lower part. Clypeus much darker than frons
                                                                    Eurysa Fieber (p. 71)
13 (10) Frons with 2 median carinae   14
-      Frons with one median carina   17
14 (13) Frons on both sides between median and lateral carinae with some sensory pits.
       Such pits present also on pronotum behind lateral carinae
                                                                 Achorotile Fieber (p. 80)
-      Frons and pronotum in adults without sensory pits (present in larvae)   15
15 (14) Carinae of frons narrow and indistinct   16
-      Carinae of frons thick and prominent, sometimes partly evanescent. Fore wings
       in brachypters black or brown with hind margin whitish
                                                               Criomorphus Curtis (p. 177)
16 (15) Body comparatively slender. Frons below with two black or dark brown spots
                                                                   Stiroma Fieber (p. 73)
-      Body robust. Frons of uniform colour, usually light *Ditropis* Kirschbaum (p. 68)
17 (13) Median keel of frons forked at least in upper third (Text-fig. 464). Vertex an-
       teriorly with 4 parallel keels at almost equal intervals
                                                               Dicranotropis Fieber (p. 152)
-      Median keel of frons forked at or near junction with vertex (Text-fig. 585). Ver-
       tex anteriorly with median keels more or less convergent   18
18 (17) Veins of fore wings with dark tubercles distinctly wider than veins   19
         -      Tubercles of fore wings, if present, narrower than veins   20
19 (18) Frons of more or less uniform light colour, below darker. Second antennal seg-
       ment with a weak keel at base (Text-fig. 203)   *Conomelus* Fieber (p. 85)
-      Frons dark with light spots. Second antennal segment without a keel
                                                               Euconomelus Haupt (p. 83)
20 (18) Fore wing with a short black or dark brown streak along claval commissure just
       proximally of apex of clavus   *Laodelphax* Fennah (p. 118)
-      Fore wing without a well-marked darker streak along commissural border   21
21 (20) Vertex about as long as broad, about pentagonal, anteriorly extending in an
       obtuse angle (Text-fig. 129). Median keels of frons weak
                                                               Delphacinus Fieber (p. 66)
-      Vertex anteriorly not extending in an angle. Median keel of frons usually strong
                                                                                   22
22 (21) Male pygofer as seen from behind twice as high as broad (Text-fig. 551). Ap-
       pendices of anal tube in male long and stout, parallel. Lateral lobes of female
       apically truncate with a shallow excision (Text-fig. 558)
                                                               Oncodelphax W. Wagner (p. 173)
-      Male pygofer not twice as high as broad. Appendices of anal tube different.
       Lateral lobes of female apically not truncate, nor excised   23
23 (22) Anterior part of body – at least mesonotum – dorsally with a more or less broad,
       whitish, longitudinal band bordering the median keel   24
-      Anterior part of body without a whitish longitudinal band on dorsum

(sometimes the median keel itself is light) 29
24 (23) Frons entirely black between its white carinae, its lateral margins evenly arched. Teeth of post-tibial calcar (except the apical one) indistinct or absent
Tyrphodelphax Vilbaste (p. 148)
– Frons between carinae not entirely black 25
25 (24) Frons narrow, almost parallel-sided, usually yellowish between the whitish carinae *Unkanodes* Fennah (p. 110)
– Frons not parallel-sided 26
26 (25) Length of body of brachypterous specimens over 3 mm, overall length of macropterous individuals over 4.2 mm. Length of fore wing in brachypters twice its width. Appendages of anal tube in male more or less parallel
Megadelphax W. Wagner (p. 112)
– Length of body of brachypters less than 3 mm, length of macropters under 4 mm. Length of fore wing in brachypters less than twice its width 27
27 (26) Vertex sordid yellow, its keels anteriorly obliterated. Frons sordid yellowish with concolorous, not or indistinctly dark-edged keels. Appendages of anal tube in male small, inconspicuous *Muirodelphax* W. Wagner (p. 144)
– Keels of wertex not obliterated, whitish with black interspaces. Keels of frons black-edged 28
28 (27) Appendages of anal tube in male usually crossed (if not crossed nearly contiguous). Fore wings of brachypters apically evenly rounded
Ribautodelphax W. Wagner (p. 202)
– Appendages of anal tube of male parallel, with a broad interspace. Fore wings of brachypters apically almost truncate *Megadelphax* W. Wagner (p. 112)
29 (23) Frons uniformly yellowish or brownish, sometimes fuscous, keels not or little lighter 34
– Frons black, fuscous or mottled, or with dark-edged keels, keels usually distinctly lighter 30
30 (29) Pygofer of male in lateral aspect with a broad and comparatively deep incision (Text-fig. 355). Frons of male between keels black, frons of female brownish yellow with blackish lines along keels. Fore wings of brachypters semi-transparent, apically truncate, index length: width = 6:5. Genital scale of female well developed, comparatively large (0.4 mm broad), black (Text-fig. 364)
Hyledelphax Vilbaste (p. 126)
– Pygofer of male in lateral aspect without a broad and deep incision. Index length: width of fore wings in brachypters at least 3:2 31
31 (30) Frons light with dark-edged carinae. Styles of male near apex produced into a short tooth projecting inwards (Text-fig. 396). Index length: width of fore wing in brachypters about 5:3 *Gravesteiniella* W. Wagner (p. 137)
– Frons in both sexes black or dark brown. Styles of male without an inwards directed tooth near apex, apically sharp-pointed or squarely truncate 32
32 (31) Frons and vertex at junction black or dark brown (also keels). Lower part of frons dark with light keels. Styles of male long, approximately parallel. Pygofer

46

of male about 1.5 times as high as broad (Text-fig. 370). First antennal segment almost as long as second segment. Frons immediately above clypeus broader than at junction with vertex. Large species *Calligypona* J. Sahlberg (p. 130)

– First antennal segment shorter. Genital segment of male not higher than broad. Smaller species 33

33 (32) Post-tibial calcar half as long as hind tarsus, on lateral margin with 17–23 small black teeth. Hind tarsus as long as proximal part of hind tibia (from basis of tibia to basis of tarsus). Index length: width of fore wing in brachypters 1.6–1.8. Styles of male slightly broadened towards apex, apical margin convex with an acute outer angle. Appendages of anal tube in male considerably remote from each other. Inner margin of lateral lobe in females only weakly concave towards base *Struebingianella* W. Wagner (p. 162)

– Post-tibial calcar usually shorter with fewer teeth on lateral margin. Styles of male tapering towards apex. Appendages of anal tube in male set close to each other. Inner margins of lateral lobes of female more distinctly concave towards base *Javesella* Fennah (p. 183)

34 (29) Males 35

– Females 44

35 (34) Apical margin of pygofer in lateral view with a broad concavity 36

– Apical margin of pygofer in lateral aspect straight or convex, or only slightly concave 37

36 (35) Pygofer below base of styles with a pointed median projection (Text-fig. 435) *Acanthodelphax* Le Quesne (p. 146)

– No median projection below base of styles *Muellerianella* W. Wagner (p. 138)

37 (35) Body light yellow, only claws, eyes, apices of styles and anal tube appendages darker. Small species, brachypters 1.9–2.6 mm, macropters (with wings) 3.1–3.6 mm *Xanthodelphax* W. Wagner (p. 166)

– Body with more or less prevalent dark markings or predominantly dark 38

38 (37) Frons narrow, index length: maximal width about 2.5. Antennal segments 1 + 2 long, reaching far beyond epistomal suture, 1st segment about twice as long as its maximal width, its apex dark. Index length: maximal width of fore wing in brachypters about 2.4 *Paradelphacodes* W. Wagner (p. 172)

– Frons broader, index length: maximal width not above 2.1 39

39 (38) Pygofer higher than broad 40

– Pygofer not higher than broad 41

40 (39) Appendages of anal tube close together. Fore wings in brachypters black or dark brown, apical margin white, index length: width about 1.6 *Florodelphax* Vilbaste (p. 156)

– Appendages of anal tube wide apart. Fore wings in brachypters not black, index length: width over 2 *Paraliburnia* Jensen-Haarup (p. 120)

41 (39) Appendages of anal tube well developed 42

– Appendages of anal tube inconspicuous 43

42 (41) Appendages of anal tube diverging. Fore wings in brachypters black or fuscous,

47

basally and in clavus pale; costa, apical margin, and proximal 2/3 of claval commissure also pale *Struebingianella* W. Wagner (p. 162)
–   Appendages of anal tube parallel. Fore wings in brachypters blackish, apical margin pale *Florodelphax* Vilbaste (p. 156)
43 (41) Abdomen above black, median line pale. Styles sickle-shaped, in lateral aspect each with a strong spine directed backwards near base (Text-fig. 499)
Kosswigianella W. Wagner (p. 160)
–   Abdomen in brachypters black with more or less broken longitudinal light streak on each side. Styles as seen from behind almost straight, without a spine near base *Muirodelphax* W. Wagner (p. 144)
44 (34) Carinae on junction vertex-frons distinct 45
–   Carinae on junction vertex-frons more or less obsolete 47
45 (44) Pronotum and scutellum pale yellowish between lateral carinae (scutellum in macropters sometimes brownish), darker beyond lateral keels. Commissural margin of fore wing (or basal half of it in macropters) whitish. Median carina of frons forked somewhat below junction with vertex. First antennal segment about twice as long as broad. Thoracal sterna light with a black or fuscous spot on the katepisterna of mesothorax. Clypeus between keels darker than frons
Muellerianella W. Wagner (p. 138)
–   Pronotum and scutellum either not uniformly pale yellowish between lateral keels or not distinctly darker beyond the latter. Commissural margin of fore wing rarely whitish. First antennal segment not twice as long as broad 46
46 (45) Basal part of lateral lobe forming acute angle (Text-fig. 494). Fore wings in brachypters brownish with apical margins broadly whitish
Florodelphax Vilbaste (p. 156)
–   Basal part of lateral lobe smoothly rounded or forming blunt obtuse or right angle. If the fore wings of brachypters are brownish, their apical margins are not broadly whitish *Javesella* Fennah (p. 183)
47 (44) Frons narrow, index length: maximal width about 2.5. Antennal segments 1 + 2 long, reaching far beyond epistomal suture, 1st segment about twice as long as its maximal width, its apex dark. Index length: maximal width of fore wing in brachypters about 2.4 *Paradelphacodes* W. Wagner (p. 172)
Frons broader, index length: maximal width at most = 2.1 48
48 (47) Index length: width of frons = 2–2.1 49
–   Frons broader, index length: width = 1.6–2.0 50
49 (48) Large species, brachypters 3.9–4.5 mm, macropters with wings 5–5.6 mm. Body yellow or light brownish yellow with or without more or less diffuse dark markings. Index length: width of fore wing in brachypters = 1.6–1.7. Lateral lobe near base with a well-marked concavity *Struebingianella* W. Wagner (p. 162)
–   Body fuscous or dirty yellow, partly black. Lateral lobe almost parallel-sided, inner margin only faintly concave *Paraliburnia* Jensen-Haarup (p. 120)
50 (48) Body light yellow or pale yellow without dark markings (except claws, eyes, ovipositor and apex of rostrum) 51

- Body not entirely light   52
51 (50) Body light yellow. Frons not very convex-sided, index length: width
1.7–2.0. Index length: width of fore wing in brachypters 1.5–1.8
Xanthodelphax W. Wagner (p. 166)
- Body pale yellow. Frons more convex-sided, index length: width of fore wing in
brachypters 1.4–1.45   *Acanthodelphax* Le Quesne (p. 146)
52 (50) Abdomen in brachypters light with well-defined black markings, in macropters
black with well-defined light markings   53
- Abdomen usually dark, markings, if present, diffuse   54
53 (52) Frons broader, convex-sided, index length: maximal width about 1.7. Small
species: brachypters 1.6–2.4 mm, macropters with wings 3.0–3.4 mm. Index
length: width of fore wing in brachypters 1.2–1.3. Inner margin of lateral lobe
convex (Text-fig. 503)   *Kosswigianella* W. Wagner (p. 160)
- Frons narrower, almost straight-sided, index length: maximal width = 1.6.
Length of brachypters 2.0–3.3 mm, macropters 3.5–4.1 mm. Index length: width
of fore wing = 1.45. Inner margin of lateral lobe faintly concave
Muirodelphax W. Wagner (p. 144)
54 (52) Frons broader, index length: maximal width about 1.65. Pronotum caudally in
both brachypters and macropters broadly whitish. Fore wings in brachypters
dark brown with apical margin broadly whitish. Index length: width of fore wing
in brachypters 1.45–1.63.   *Florodelphax* Vilbaste (p. 156)
- Frons narrower, index length: maximal width at least 1.75. Fore wings of
brachypters may be dark but then their apical margin is not broadly whitish
(only the marginal vein being sometimes light)   *Javesella* Fennah (p. 183)

SUBFAMILY KELISIINAE

First tarsal segment of hind tibiae with 4 + 2 spines. Body comparatively slender.
Usually with a roundish black spot on each of the genae. Clypeus not darker than frons.
Aedeagus often with a rudimentary theca and with a rudimentary membranous apical
part. Anal segment of male often with filamentous appendages.

Genus *Kelisia* Fieber, 1866

Kelisia Fieber, 1866b: 519.
Type-species: *Delphax guttula* Germar, 1818, by monotypy.

Vertex not twice as long as width at middle, apically fairly rounded. Median carina of
frons obsolescent on its upper apex. Wings long and narrow, fore wings even in
brachypters longer than abdomen or at least as long as the latter. Marginal teeth of
post-tibial calcar few in number (5–10), apical tooth as large as marginal teeth. Body

fairly slender. Theca of aedeagus virtually absent or transformed into a curved horn. In Denmark and Fennoscandia 6 species, all of them wing dimorphic.

Key to species of *Kelisia*

1 Black spot on gena, if present, small, occupying only half of the space between anterior and median keels. Body of an almost uniform light yellow colour
9. *pallidula* (Boheman)
– The black spot on each gena occupies at least the entire space between anterior and median keels (Text-figs. 3, 81)   2
2 (1) Anterior and median tibiae each with two black longitudinal streaks, one along exterior margin, and one along interior margin   3
– Anterior and median tibiae without distinct longitudinal streaks   5
3 (2) Anal tube in male with one unsymmetrical, curved, horn-shaped appendage (Text-fig. 78)   12. *monoceros* Ribaut
– Anal tube in male with two symmetrical, weakly S-shaped appendages (Text-fig. 73)   4
4 (3) Smaller, total length of male 2.8–3.6 mm, of female 3.0–4.0 mm. Length of aedeagus 0.65–0.78 mm. Index length of aedeagus: total length = 1: 3.7–4.8. Brachypters with fore wings apically strongly narrowed   11. *ribauti* W. Wagner
– Larger, total length of males 3.1–4.0 mm, of females 3.8–4.3 mm. Length of aedeagus 0.57–0.68 mm. Index length of aedeagus: total length 1: 5.0–6.0. Brachypters with fore wings a little shorter and with the same shape as those of macropters   10. *sabulicola* W. Wagner
5 (2) Black spot on gena reaching posterior keel (Text-fig. 81)   13. *guttula* (Germar)
– Black spot on gena at most extending half the distance between median and posterior keel, normally not beyond median keel (Text-fig. 3)
14. *vittipennis* (J. Sahlberg)

9. *Kelisia pallidula* (Boheman, 1847)
Text-figs. 62–68.

Delphax pallidula Boheman, 1847b: 265.
Delphax raniceps Boheman, 1847a: 52.

Brachypters pale yellowish, fore wings about as long as abdomen, pale with con-

---

Text-figs. 62–68. *Kelisia pallidula* (Boheman). – 62: male pygofer from below; 63: male pygofer from the right; 64: male anal tube from the right; 65: right genital style; 66: aedeagus; 67: apex of aedeagus; 68: caudal part of female abdomen from below. Scale: 0.1 mm.
Text-figs. 69, 70. *Kelisia sabulicola* W. Wagner. – 69: aedeagus from the right; 70: caudal part of female abdomen from below. Scale: 0.1 mm.

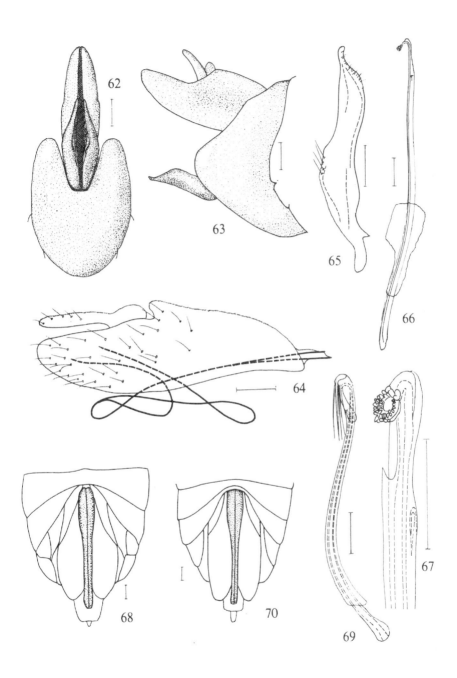

62

63

65

64

66

68

70

67

69

51

colorous or darker veins. Median keel of frons obliterated on junction with vertex, always simple. Genae with or without a more or less distinct dark spot. Also prosternum on each side with or without a dark spot. This is also valid for the macropterous form which is suggestive of a strongly faded *vittipennis* or *ribauti*. Thus, pro- and mesonotum are often dorsolaterally dirty brownish. Abdomen often partly darkened, veins of fore wings partly fuscous. Fore wings of macropters considerably longer than abdomen, apically with a diffuse cuneiform dark spot tapering towards basis. Male pygofer as in Text-figs. 62 and 63, anal tube as in Text-fig. 64, style as in Text-fig. 65, aedeagus as in Text-figs. 66 and 67, posterior part of female abdomen from below as in Text-fig. 68. Length of brachypters 2.3–3.3 mm, of macropters 3.3–3.7 mm.

Distribution. Fairly common in Denmark, found in most districts. – Widespread and comparatively common in the south of Sweden up to Upl. – Norway: found in AK, VAy, and TEy. – Comparatively rare in East Fennoscandia, recorded from Al, Ab, N, St, and Ta. – North and Central Eûrope, also in Kazakhstan, Uzbekistan, and Mongolia.

Biology. Among *Carices* in moors, bogs, moist meadows etc., adults in July–September.

10. *Kelisia sabulicola* W. Wagner, 1952
   Text-figs. 69, 70.

Kelisia sabulicola W. Wagner, 1952: 3.

Carinae of head distinct, median carina of frons simple. Genae each with a roundish black spot reaching from anterior to median keel of gena. Vertex pale yellowish. A small black spot on each side of prothorax. Pro- and mesonotum pale yellow, on each side laterad of lateral keels with a broad fuscous longitudinal band. The latter is sometimes strongly reduced. Abdomen dorsally mainly, ventrally partly, black. Anterior and median tibiae, often also posterior tibiae, with a longitudinal black streak on exterior and interior margins. Fore wings hyaline with a blackish oblong spot at apex. Anterior and median femora with two longitudinal black streaks. Fore wings of the brachypterous form a little shorter than those of macropters but apically broadly rounded like the latter. In both forms the fore wings are longer than abdomen. Anal tube of male with two symmetrical appendages (as in *ribauti*). Aedeagus as in Text-fig. 69, posterior part of female abdomen from below as in Text-fig. 70. Total length (with wings) of male 3.1–4.0 mm, of females 3.8–4.3 mm.

Distribution. So far not found in Denmark, nor in Norway. – Reliable Swedish records from Sk., Bl., Hall., and Öl. – East Fennoscandia: N: Tvärminne (Albrecht, 1977). – Great Britain, Ireland, Germany, Netherlands, Poland.

Biology. On sand dunes, according to Wagner (1952) monophagous on *Carex arenaria*. Adults in August–October.

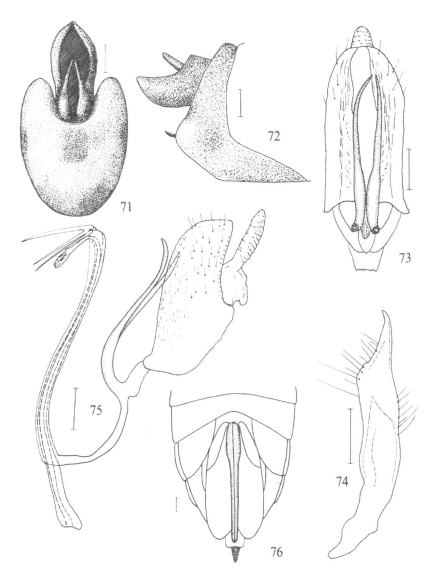

Text-figs. 71-76. *Kelisia ribauti* W. Wagner. – 71: male pygofer from below; 72: male pygofer from the right; 73: male anal tube from below; 74: left genital style from outside; 75: aedeagus and anal tube from the right; 76: caudal part of female abdomen from below. Scale: 0.1 mm.

11. *Kelisia ribauti* W. Wagner, 1938
Text-figs. 71–76.

Kelisia ribauti W. Wagner, 1938: 12.
Kelisia guttula Ribaut, 1934: 292, (nec Germar, 1818).

Very like *sabulicola*, but fore wings of the brachypterous form apically narrowed. Fore wings in both forms longer than abdomen. Overall length of males 2.8–3.6 mm, females 3.0–4.0 mm. Male pygofer as in Text-figs. 71, 72, anal tube of male as in Text-figs. 73 and 75, style as in Text-fig. 74, aedeagus as in Text-fig. 75, posterior part of female abdomen from below as in Text-fig. 76.

Distribution. Common and widespread in Denmark, Sweden (up to Jmt.), and Finland (up to ObN). – In Norway so far only in AK, VAy, Ry, and Nsy. – Europe, N. Africa, Altai, Georgia, Kazakhstan, Kirghizia, Tadzhikistan, Uzbekistan.

Biology. On *Carices* in damp meadows etc., adults from July on. Hibernation takes place in the egg stage (Müller, 1957).

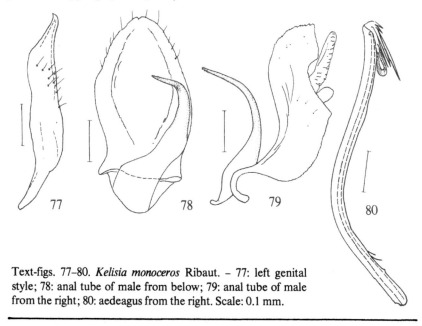

Text-figs. 77–80. *Kelisia monoceros* Ribaut. – 77: left genital style; 78: anal tube of male from below; 79: anal tube of male from the right; 80: aedeagus from the right. Scale: 0.1 mm.

Text-figs. 81–88. *Kelisia guttula* (Germar). – 81: head and prothorax from the left; 82: male pygofer from below; 83: male pygofer from the right; 84: male anal tube from below; 85: male anal tube from the left; 86: left genital style; 87: apical part of aedeagus with detail in higher magnification; 88: caudal part of female abdomen from below. Scale: 0.1 mm.

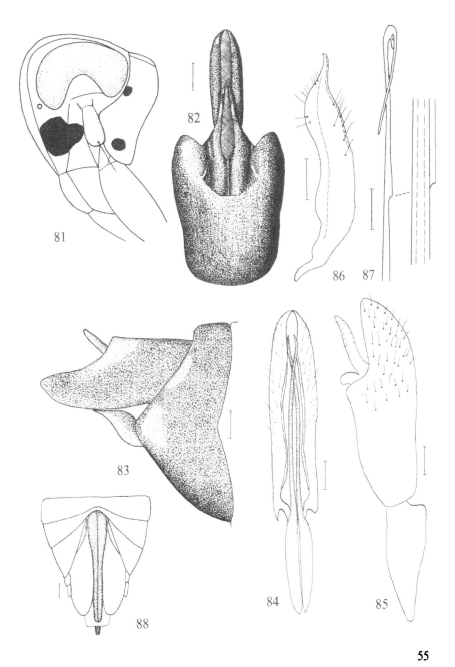

81

82

86 87

83

84 85

88

55

12. **Kelisia monoceros** Ribaut, 1934
Text-figs. 77–80.

Kelisia monoceros Ribaut, 1934: 293.

Very like *ribauti*, differing by characters given in the key. Styles as in Text-fig. 77, male anal tube as in Text-figs. 78 and 79, aedeagus as in Text-fig. 80. Total length 2.5–3.5 mm.

Distribution. So far not found in Denmark. – In Sweden recorded from Bl., Sm., Öl., Gtl., Vg., Upl., and Vstm. – Warloe found *Kelisia monoceros* in Norway, Bø: Drammen and Ringerike. – Comparatively rare in East Fennoscandia, found in Al, Ab, and Sa. – Austria, Czechoslovakia, France, BRD, DDR, Italy, Poland, Romania, East Baltic, and Central Russia.

Biology. Apparently a thermophilous species living on dry, sunny localities, adults in July and August.

13. **Kelisia guttula** (Germar, 1818)
Plate-fig. 3, text-figs. 81–88.

Delphax guttula Germar, 1818: 216.
Kelisia pascuorum Ribaut, 1934: 294.

Like *sabulicola* but without black streaks on tibiae. Black spot on gena reaching considerably caudad of median keel of gena (Text-fig. 81). The lateral longitudinal band on pronotum is often reduced. Male pygofer as in Text-figs. 82 and 83, male anal tube as in Text-figs. 84 and 85, style as in Text-fig. 86. Aedeagus basally widely curved (as in *vittipennis*, Text-fig. 94), distally straight up to apex which wears only one long needle-shaped appendage. Aedeagus with a longitudinal carina proximally ending in a rectangular corner (Text-fig. 87). Apical part of female abdomen in ventral aspect as in Text-fig. 88. Overall length 2.75–3.5 mm.

Distribution. Scarce in Denmark, found only in SJ and B. – Common in southern Sweden, Sk. – Ång. – In Norway so far found in Ø, TEy, Ry, and HOi. – Fairly rare in East Fennoscandia, recorded from Al, Ab, N, Ta, Vib, and Kr. – Widespread in Europe, also in North Africa, Azerbaijan, Tadzhikistan, and middle Siberia.

Biology. On sedges in both dry and wet localities, adults in July–September.

14. **Kelisia vittipennis** (J. Sahlberg, 1868)
Text-figs. 3, 89–95.

Delphax vittipennis J. Sahlberg, 1868: 187.
Stenocarenus guttuliferus J. Sahlberg, 1871: 416 (nec Delphax guttulifera Kirschbaum, 1868).

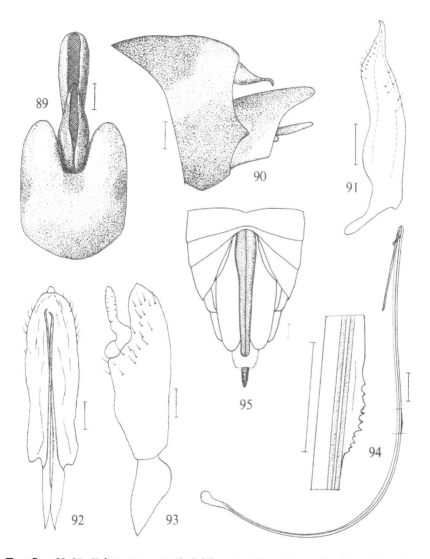

Text-figs. 89–95. *Kelisia vittipennis* (J. Sahlberg). – 89: male pygofer from below; 90: male pygofer from the right; 91: left genital style; 92: male anal tube from below; 93: male anal tube from the left; 94: aedeagus from the left, with detail in higher magnification; 95: caudal part of female abdomen from below. Scale: 0.1 mm.

As *sabulicola* but usually larger and without dark longitudinal streaks on tibiae. Dark pigmentation of dorsum often more extended than in related species. In macropters the black longitudinal band of the fore wing is often broad and distinct to the very base of the wing. Male pygofer as in Text-figs. 89 and 90, style as in Text-fig. 91, anal tube of male as in Text-figs. 92 and 93, aedeagus (Text-fig. 94) as in *guttula*, but its longitudinal carina on its broadest part with 8–10 minute teeth, proximally gradually tapering. Venter of posterior part of female abdomen as in Text-fig. 95. Overall length 3–4 mm.

Distribution. Fairly common in Denmark, found in EJ, NWJ, LFM, SZ, and NEZ. – Not uncommon in Sweden, found in most provinces from Sk. to Nb. – I have seen specimens from AK, VAy, Ry, and Ri in Norway. – Tolerably common in southern and central East Fennoscandia, found in Ab, N, Ta, Sa, Oa, Kb, Om, Kr. – Widespread in Europe, also in Algeria and middle Siberia.

Biology. In bogs with *Eriophorum vaginatum* (Linnavuori, 1952a). Adults from August on. Hibernation in the egg stage (in Central Europe, Müller, 1957).

Genus *Anakelisia* W. Wagner, 1963

Anakelisia W. Wagner, 1963: 165.
Type-species: *Ditropis fasciata* Kirschbaum, 1868, by original designation.

General aspect and shape of body as in *Kelisia*, but aedeagus with a distinct theca, without terminal needle-like appendages. In Denmark and Fennoscandia two species, both wing dimorphic.

Key to species of *Anakelisia*

1 Black spot on genae, if present, occupying at most half distance between anterior and median keel. Large species, overall length 3.2–5.4 mm. Anal tube of male light-coloured, not prolonged caudally (Text-figs. 96–99). Saw-case of female short, caudally not extending beyond apical end of pygofer (Text-fig. 103)
15. *fasciata* (Kirschbaum)
– Black spot on genae extending whole distance between anterior and median keels. Small species, length 1.3–3.2 mm. Anal tube of male caudally strongly prolonged, partly black (Text-figs. 104, 105). Saw-case of female long, reaching beyond apex of pygofer (Text-fig. 110)
16. *perspicillata* (Boheman)

Text-figs. 96–103. *Anakelisia fasciata* (Kirschbaum). – 96: male pygofer from behind; 97: male pygofer from the right; 98: male anal tube from below; 99: male anal tube from the left; 100: left genital style from outside; 101: aedeagus from behind; 102: aedeagus from the left; 103: caudal part of female abdomen from below. Scale: 0.1 mm.

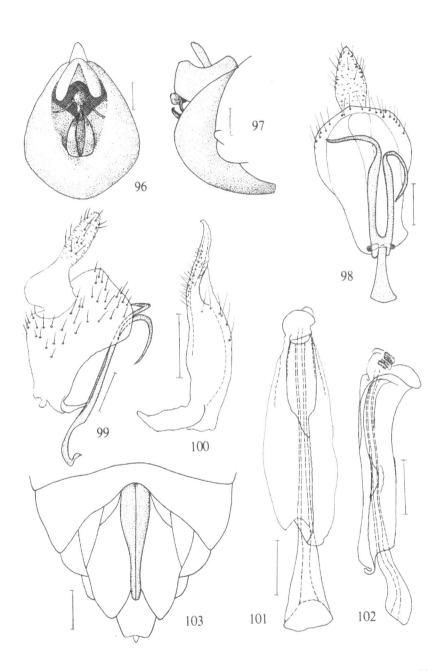

59

15. *Anakelisia fasciata* (Kirschbaum, 1868)
Text-figs. 96–103.

Ditropis fasciata Kirschbaum, 1868: 42.
Delphax scotti Scott, 1870: 25.

Yellowish, comparatively strongly built. Usually with a small black spot on each gena and another on each side of prothorax. Median keel of frons often partly doubled or obsolete. Fore wings of brachypters slightly longer than abdomen, those of macropters about one-third longer than abdomen. Sometimes they are pale without dark markings, usually there is at least one blackish spot at apex, often also a dark spot in apex of clavus and/or one at the wing basis. In the female the corium is often partly blackish. Pygofer of male as in Text-figs. 96 and 97, anal segment of male as in Text-figs. 98 and 99, style as in Text-fig. 100, aedeagus as in Text-figs. 101 and 102, venter of apical part of female abdomen as in Text-fig. 103. Overall length of brachypters 3.2–4.2 mm, of macropters 4.8–5.2 mm.

Distribution. Scarce in Denmark, found in F, SZ, and NEZ. – Rare in Sweden, only found in Öl., Halltorp (H. Andersson & R. Danielsson), Upl., Djursholm, lake Ösbysjön (C. H. Lindroth, Ossiannilsson), and in the vicinity of Örebro (A. Jansson). – So far not found in Norway, nor in East Fennoscandia. – Austria, Czechoslovakia, France, BRD, DDR, England, Hungary, Poland, Romania.

Biology. On *Carex* (*riparia,* according to Müller, 1951) at lake-shores etc. Adults in August–October. Hibernation in the egg stage or occasionally by adult females (Müller, 1957).

16. *Anakelisia perspicillata* (Boheman, 1845)
Text-figs. 104–110.

Delphax perspicillata Boheman, 1845: 164.

Sordid pale yellowish. Median carina of frons above indistinct, at middle sometimes doubled or broadened. Face brownish yellow. Vertex, pronotum, and mesonotum brownish yellow or dirty pale yellowish without markings. Genae each with a roundish black spot not going beyond their median keel. Abdomen partly fuscous. Prothorax laterally with a black spot. Fore wings of brachypters about as long as abdomen, without dark markings. The macropterous form is extremely rare. Fore wings of macropters with fuscous veins and an oblong blackish spot at apex. Male pygofer as in Text-figs. 104 and 105. Anal tube of male (Text-figs. 106, 107) large, partly black, with two almost straight needle-like appendages. Styles as in Text-fig. 108. Theca of aedeagus (Text-fig. 109) with two strong pointed appendages distally of middle. Posterior part of female abdomen from beneath as in Text-fig. 110. Sawcase long, extending beyond apex of pygofer. Length of brachypters 1.3–2.2 mm, of macropters 3.1–3.3 mm.

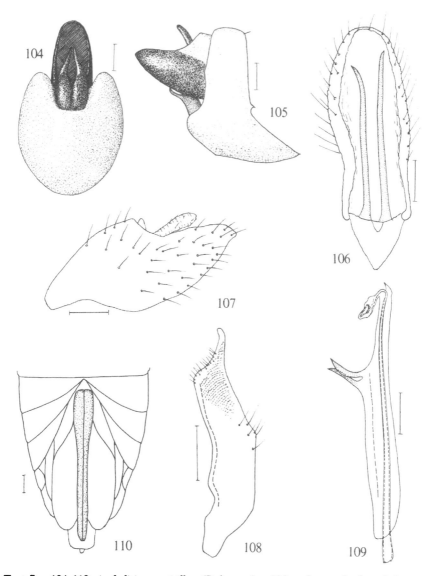

Text-figs. 104–110. *Anakelisia perspicillata* (Boheman). – 104: male pygofer from below; 105: male pygofer from the right; 106: male anal tube from below; 107: male anal tube from the left; 108: left genital style; 109: aedeagus from the right; 110: caudal part of female abdomen from below. Scale: 0.1 mm.

Distribution. Rare in Denmark, only found in EJ: Tebbestrup bakker 6.V.1878 (O. Jacobsen), and in NEZ: Tisvilde, September (Schlick). – Scarce in Sweden, Sm. – Upl. and Nrk. – Not in Norway, nor in East Fennoscandia. – Widespread in Europe, also found in Mongolia and Siberia.

Biology. Preferably on dry meadows, slopes etc. On *Carex*. Hibernation (in Central Europe) in the egg stage (Schiemenz, 1969).

SUBFAMILY STENOCRANINAE

Vertex more or less narrow. Frons of last instar larva parallel-sided. Frons of adults narrow, not conspicuously broadened in lower part. Body comparatively slender. Teeth of post-tibial calcar broadened into parallel-sided plates. Sawcase of females broad, shield-shaped, ventrally with numerous wax-glands.

Genus *Stenocranus* Fieber, 1866

Stenocranus Fieber, 1866b: 519.

Type-species: *Fulgora minuta* Fabricius, 1787, by subsequent designation.

Frons elongate, narrowing between eyes. Vertex elongate, apically more or less protruding in front of eyes. Median carina of frons simple. Wings longer than abdomen, fore wings very elongate. Aedeagus partly enclosed in a theca. Saw-case of female very broad, conceiling ventral part of pygofer (Text-figs. 119, 128). In Denmark and Fennoscandia two species.

Key to species of *Stenocranus*

1 In lateral aspect, distance between compound eye and apical margin of frons about twice as long as shortest distance between eye and margin of frons (Text-fig. 111). Vertex and frons between carinae with two brownish (not black) longitudinal streaks. Surface of fore wing between veins finely transversely wrinkled

17. *minutus* (Fabricius)

– In lateral aspect, distance between eye and apical point of margin of frons about one and a half times as long as shortest distance between eye and margin of frons (Text-fig. 120). Vertex and frons between keels with black longitudinal lines. Fore wing membrane between veins not distinctly transversely wrinkled

18. *major* (Kirschbaum)

Text-figs. 111–119. *Stenocranus minutus* (Fabricius). 111: head and pronotum from the left; 112: male pygofer from behind; 113: male pygofer from the right; 114: male anal tube from behind; 115: male anal tube from the left; 116: genital style; 117: aedeagus from the left; 118: aedeagus, ventral aspect; 119: caudal part of female abdomen from below. Scale: 0.5 mm for 119, 1 mm for 111, 0.1 mm for the rest.

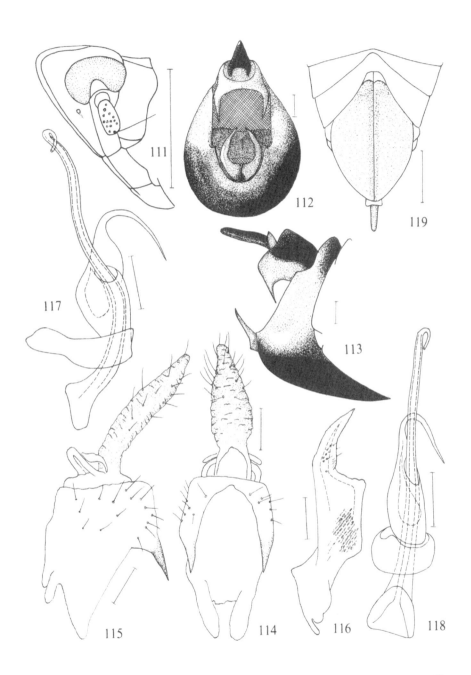

111

112

119

117

113

115 114 116 118

63

17. *Stenocranus minutus* (Fabricius, 1787)
Plate-fig. 27, text-figs. 111–119.

Fulgora minuta Fabricius, 1787: 262.
Delphax lineola Germar, 1822: pl. 19.

Elongate, pale yellow. Vertex, pronotum and mesonotum with a whitish longitudinal streak on and along median carina. Laterally of this streak there is a more or less distinct, orange-coloured, longitudinal band on each side. Region immediately laterally of side-keels of pro- and mesonotum pale. On the fore wing there is a black longitudinal band varying in extent. In most specimens this band is well developed only in apical half of the wing where it is broader towards apex; veins only being dark in basal part of the wing. These dark markings may be entirely absent, but on the other hand most of the inner half of the fore wing may be dark. The species is wing-dimorphous, but even in brachypters the wings are longer than abdomen; in the macropterous form they are still considerably longer. Fore wings between veins transversely wrinkled. Femora on underside with 2 black longitudinal streaks, tibiae laterally with a black longitudinal streak. Abdomen partly black. Male pygofer as in Text-figs. 112, 113. Anal segment of male (Text-figs. 114, 115) long, black. Styles (Text-fig. 116) slender, together resembling a pair of forceps (or the "forceps" of the male of the common earwig). Aedeagus as in Text-figs. 117 and 118. Posterior part of female abdomen as in Text-fig. 119. Overall length 4.5–5.8 mm.

Distribution. Common and widespread in Denmark. – In Sweden, *Stenocranus minutus* appears to have two centres of distribution. It is comparatively common in Scania and in Sdm. and Upl. but apparently rare in the territory between these centres (Sk., Bl., Sm., Öl., Dlsl., Sdm., Upl., Vstm.). – In Norway found only in or near Oslo (AK, Holgersen, 1946). – East Fennoscandia: Al, Eckerö (Håkan Lindberg), Ab: Pojo, Spakanäs (Albrecht, 1977). – Widespread in Europe, also in North Africa, Azerbaijan, Kazakhstan, Uzbekistan, and Kirghizia.

Biology. On grasses in meadows, dunes, sandy plains, marshes, and forests. Host-plant: *Dactylis glomerata* (Müller, 1942). Adults September–June, hibernation takes place in the adult stage. The diapause of *S. minutus* has been studied by Müller (1957).

18. *Stenocranus major* (Kirschbaum, 1868)
Text-figs. 120–128.

Text-figs. 120–128. *Stenocranus major* (Kirschbaum). – 120: head and pronotum from the left; 121: male pygofer from behind; 122: male pygofer from the right; 123: male anal tube from behind; 124: male anal tube from the left; 125: genital style; 126: aedeagus from the left; 127: aedeagus, ventral aspect; 128: caudal part of female abdomen from below: Scale: 1 mm for 120, 0.5 mm for 128, 0.1 mm for the rest.

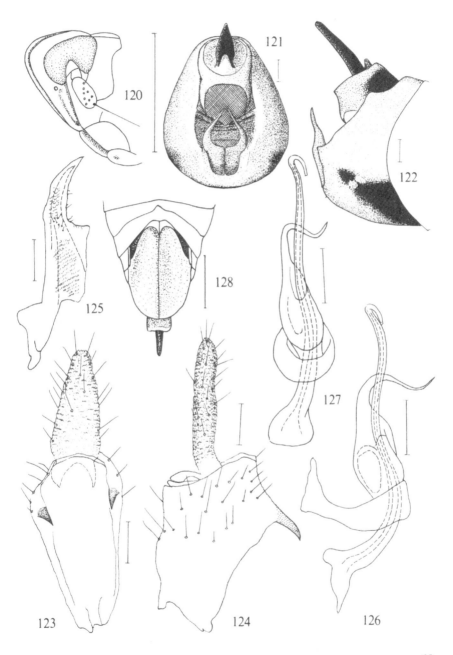

120

121

122

125

128

127

123

124

126

Delphax major Kirschbaum, 1868: 21.
Stenocranus fuscovittatus Jensen-Haarup, 1920: 39, nec Stål, 1858.

Like *minutus* from which the present species can be distinguished by characters given in the key. *Stenocranus major* is also a little larger with an overall length of 5.4–6.7 mm (according to Le Quesne, 1960). Male pygofer as in Text-figs. 121 and 122, anal segment of male as in Text-figs. 123 and 124, style as in Text-fig. 125, aedeagus as in Text-figs. 126 and 127, venter of posterior part of female abdomen as in Text-fig. 128.

Distribution. Rare in Denmark, found by several collectors in NEZ: Grib skov. Also in B: Ypnasted 20.VIII.1976 and Ekkodalen 8.IX.1977 (L. Trolle). – Rare also in Sweden, found in Sk.: Trolleholm 21.V.1936, 1 ♀ (Ossiannilsson), and in Öl.: Halltorps hage 28–31.VIII.1976, 1 ♂, 1 ♀ (H. Andersson & R. Danielsson). – Not found in Norway, nor in East Fennoscandia. – Recorded from Austria, Czechoslovakia, BRD, DDR, England, Ireland, Netherlands, Poland, Yugoslavia, Italy.

Biology. On *Phalaris arundinacea* (Le Quesne, 1960, L. Trolle in litt.). Hibernation takes place in the adult stage (Müller, 1957). Occasionally acting as a pest in rice-fields in Italy (Olmi, 1969).

SUBFAMILY STIROMINAE

First tarsal segment of hind tibiae with 5 + 2 or 6 + 2 spines. Aedeagus without rudiments of theca and without membranous apical part. Carinae of frons weak. Median carina either simple, or double and parallel, or obsolescent, but not forked. Frons above broadened or as broad as below, not narrower between upper corners of compound eyes. Cuticula of frons often glossy or strongly shining. Clypeus often uniformly darkened.

Genus *Delphacinus* Fieber, 1866

Delphacinus Fieber, 1866b: 520.
Type-species: *Delphax mesomelas* Boheman, 1849, by monotypy.

Vertex anteriorly prolonged in an obtuse angle. Frons broad, narrowest near clypeus, with a single median carina becoming weakened at the very transition to vertex. Carinae of vertex distinct but not very sharp. Wing dimorphous. Pro- and mesonotum each with 3 keels, partly less distinct in the macropterous form. In Europe only one species.

Text-figs. 129–138. *Delphacinus mesomelas* (Boheman). – 129: head from above; 130: male pygofer from behind; 131: male pygofer from the right; 132: male anal tube from behind; 133: male anal tube from the left; 134: genital style; 135: aedeagus, ventral aspect; 136: aedeagus from the left; 137: aedeagus from the right; 138: caudal part of female abdomen from below. Scale: 0.25 mm for 138, 0.1 mm for the rest.

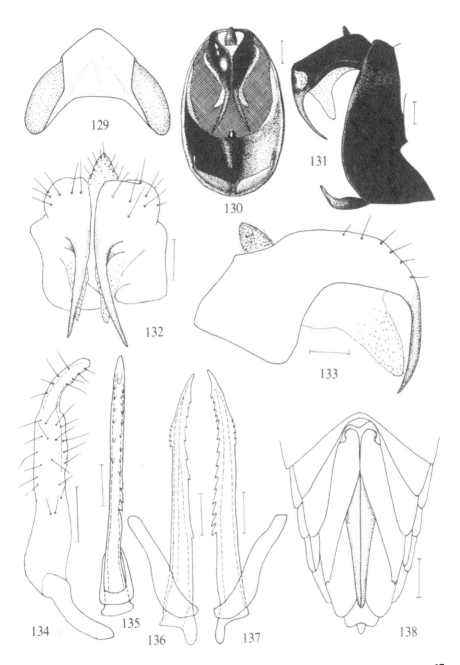

129

130

131

132

133

134   135   136   137   138

67

19. *Delphacinus mesomelas* (Boheman, 1849)
Plate-fig. 7, text-figs. 129-138.

Delphax mesomelas Boheman, 1849: 257.

Vertex pentagonal (Text-fig. 129). Body of male ventrally black under a line extending from the epistomal suture caudad along the costal margin of the resting fore wings. Dorsally of that line, head and thorax are yellowish. Fore wings of the brachypterous male (Plate-fig. 7) whitish-hyaline, those of macropters transparent with veins basally pale, a little darker towards apex. Abdomen above shining black except the yellowish two first segments and a whitish transverse band immediately in front of pygofer. The female is entirely whitish-yellow except claws, apices of tibial spines and apical parts of the fore wing veins in the macropterous form. Pygofer of male as in Text-figs. 130 and 131. Anal tube of male (Text-figs. 132, 133) large, with two long curved pointed teeth and two shorter but broader membranous appendages. Styles as in Text-fig. 134. Aedeagus (Text-figs. 135-137) almost straight, slender. Ventral aspect of posterior part of female abdomen as in Text-fig. 138. Lateral lobes of female broad, basally contiguous, concealing basis of saw-case, apically each with a small notch. Overall length of brachypters 2-3 mm, of macropters 3.5-4 mm.

Distribution. Fairly common in Denmark, found in EJ, WJ, NEJ, F, LFM, NEZ, and B. – Common in the southern part of Sweden, Sk. – Upl. – So far not found in Norway. – Scarce in southern Finland (Ab). – Widespread in Western, Central, and Eastern Europe, also found in Kazakhstan. Not in the Mediterranean area.

Biology. On grasses in meadows and dry heaths, adults in June–August. Hibernation in larval stages (Müller, 1957).

Genus *Ditropis* Kirschbaum, 1868

Ditropis Kirschbaum, 1868: 11.

Type-species: *Delphax pteridis* Spinola, 1839, by subsequent designation.

Frons with two thin carinae which become almost extinct upvards, uniformly yellowish. Pronotum almost as long as vertex. Denticles of post-tibial calcar well developed. Only one species.

20. *Ditropis pteridis* (Spinola, 1839)
Plate-figs. 5, 26, text-figs. 139-147.

Delphax pteridis Spinola, 1839: 334.
Delphax pteridis Boheman, 1852b: 115.

Legs comparatively long. Fore wings of brachypters apically truncate, reaching about to hind border of 4th abdominal segment. Brachypterous male (Plate-fig. 5): black,

68

head and legs yellow, pronotum yellow, laterally black, mesonotum, fore wings, and abdomen shining black or fuscous, anal tube and its appendages brownish yellow. Brachypterous female (Plate-fig. 26): head, thorax and legs yellow or brownish yellow, fore wings brownish yellow or fuscous, abdomen fuscous or black with yellowish margins. Fore wings of macropters transparent, basally brownish, apically dirty whitish, veins brownish. Male pygofer as in Text-figs. 139, 140, anal tube of male as in Text-figs. 141, 142, styles (Text-fig. 143) basally each with a strong backwards directed pointed process. Aedeagus as in Text-figs. 144–146, venter of posterior part of female abdomen as in Text-fig. 147. Overall length of brachypters 2.3–3.5 mm, macropters 4.4–4.7 mm.

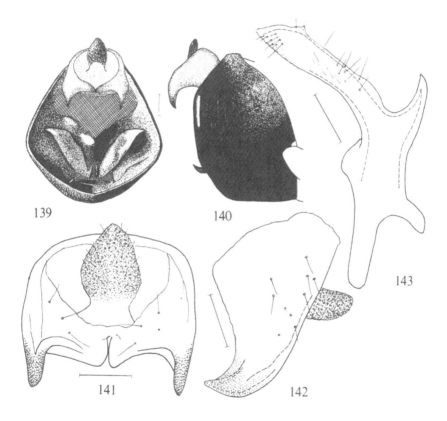

Text-figs. 139–143. *Ditropis pteridis* (Spinola). – 139: male pygofer from behind; 140: male pygofer from the right; 141: male anal tube from behind; 142: male anal tube from the left; 143: genital style. Scale: 0.1 mm.

Distribution. Scarce in Denmark, found in SJ, EJ, NWJ, and F. – Rare but locally abundant in Sweden: Sk., Bl., Sm., Gtl. – In Norway only found by Warloe in AAy: Risør. – So far not found in East Fennoscandia. – Widespread in Europe, also in USSR (Georgia).

Biology. On *Pteridium aquilinum*, adults in May–July. Hibernation in larval stages (Müller, 1957).

Text-figs. 144–147. *Ditropis pteridis* (Spinola). – 144: aedeagus, ventral aspect; 145: aedeagus from the left; 146: aedeagus from the right; 147: caudal part of female abdomen from below. Scale: 0.25 mm for 147, 0.1 mm for the rest.

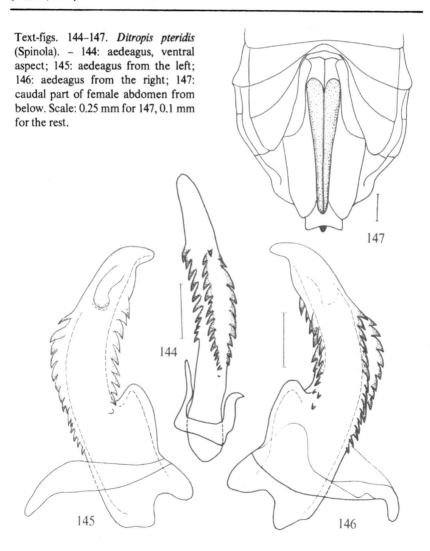

Genus *Eurysa* Fieber, 1866

Eurysa Fieber, 1866b: 520.
Type-species: *Delphax lineata* Perris, 1857, by subsequent designation.

Vertex shorter than broad, its fore border almost straight or faintly convex. Frons with a simple median carina upwards obsolete, distinct in its lower part. Keels of pro- and mesonotum partly obsolete or weak, lateral keels of pronotum posteriorly curved outwards. Antennae short. Denticles of post-tibial calcar showing tendency of reduction. Colour of clypeus more or less darkened. Abdominal segments VI and VII in larval instars on each side with a median and two lateral sensory pits. In Northern Europe one species.

21. *Eurysa lineata* (Perris, 1857)
Plate-fig. 6, text-figs. 148–154.

Delphax lineata Perris, 1857: 171.

Yellowish white or sordid yellow, fairly shining, surface of fore body wrinkled – punctured. Frons brownish with light keels; between these on each side four light spots in a vertical row and two additional light spots in a vertical row and two additional light spots near each keel. Clypeus fuscous. Pronotum and scutellum whitish yellow with four broad brownish longitudinal bands, or brownish with three light longitudinal streaks. Length of fore wings of brachypters 1.5 times width. Fore wings whitish yellow to brownish yellow. Abdomen brownish with whitish yellow median line and several lateral rows of light spots. Macropters darker with more indistinct markings, fore wings with a brownish streak along commissural margin. Male pygofer as in Text-figs. 148 and 149, anal tube of male as in Text-fig. 150, styles as in Text-figs. 151, 152, aedeagus as in Text-fig. 153. Venter of female abdomen as in Text-fig. 154, genital scale large, black. Overall length of brachypters 2.4–3.4 mm, macropters 3.6–4.4 mm.

Distribution. Very rare in Sweden: Sk., Brunnby, Kullen 16.VII.1952 (Bo Tjeder); Gtl., Klinte (P. H. Lindberg), Visby 8.VII.1952 (Ossiannilsson), Stora Karlsö 10.VII.1956 (O. Lundblad). – Not found in Denmark, Norway, and East Fennoscandia. – Otherwise widespread in Europe, also in Georgia, Jordan, Syria, Uzbekistan, and Mongolia.

Biology. On more or less dry grass meadows. Strübing (1956) mentions *Poa nemoralis* as a probable natural host-plant. Hibernation takes place in larval instars (Strübing in Müller, 1957).

Genus *Eurysula* Vilbaste, 1968

Eurysula Vilbaste, 1968: 70.
Type-species: *Eurysa lurida* Fieber, 1866, by original designation.

71

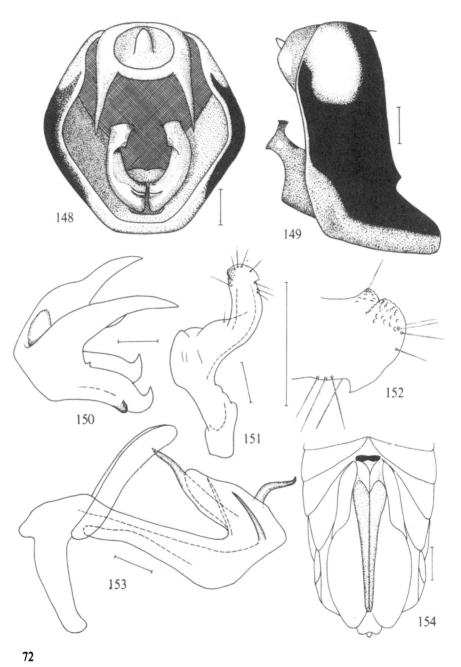

148

149

150

151

152

153

154

72

As *Eurysa*, but median carina of frons indistinct in its entire length. Abdominal segments VI and VII of larval instars each with one medial and three lateral sensory pits on each side. In North Europe one species.

22. *Eurysula lurida* (Fieber, 1866)
Text-figs. 155–162.

Eurysa lurida Fieber, 1866b: 523.
Atropis laevifrons J. Sahlberg, 1871: 484.

Vertex with three small shallow pits, one of these situated a little in front of the others. Antennae short. Lateral keels of pronotum caudally shortened, median keel missing. Mesonotum without keels. Fore wings of brachypters a little longer than their maximal width, apically rather abruptly cut off. In the male, abdomen and fore wings (of brachypters) and parts of underside of thorax are shining black, the scutellum may be black or brownish, the rest of body being brownish. Male pygofer as in Text-figs. 155 and 156, anal tube of male as in Text-fig. 157, style as in Text-fig. 158, aedeagus as in Text-figs. 159–161, venter of posterior part of female abdomen as in Text-fig. 162. Overall length of brachypters 2.2–3 mm, of macropters 3.9–4.1 mm.

Distribution. Rare in Denmark, only found in NEJ: Tranum klit 29.VI.1956 by O. Bøggild. – Rare also in Sweden: Bl.: Jämjö, Torhamn 2.VII.1963 (N. Gyllensvärd), Karlskrona, Gullberna 24.VI.1970 (Gyllensvärd); Gtl. (R. Remane in litt.); Jmt., Bispgården 6.VIII.1964 (A. Sundholm). – Very rare in Norway, only found in AAy: Risør 14.VI.1903 (Warloe). – Scarce and sporadic in East Fennoscandia, found in Al, Ab, N, Tb, Sb, Kb, and Kr. – Widespread in Europe, also found in Kazakhstan.

Biology. According to Linnavuori (1952) on *Calamagrostis epigeios*. Strübing (1955) found and reared *E. lurida* on *Calamagrostis canescens*. Hibernation takes place in larval stages (Strübing in Müller, 1957). Our adults were found in June–August.

Genus *Stiroma* Fieber, 1866

Stiroma Fieber, 1866b: 521.
Type-species: *Stiroma affinis* Fieber, 1866, by subsequent designation.

Frons vaulted with two thin median carinae indistinct in their upper part. Vertex broader than long, approximately parallel-sided, a little longer than pronotum medially.

Text-figs. 148–154. *Eurysa lineata* (Perris). – 148: male pygofer from behind; 149: male pygofer from the right; 150: male anal tube from the left; 151: right genital style; 152: apex of right genital style; 153: aedeagus from the left; 154: caudal part of female abdomen from below. Scale: 0.25 mm for 154, 0.1 mm for the rest.

73

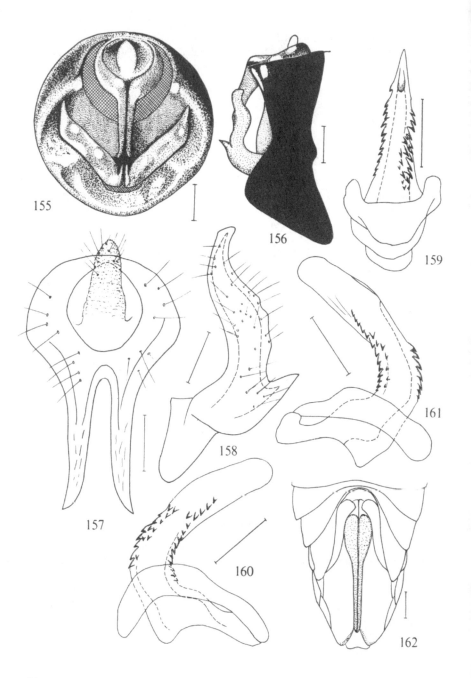

155

156

159

157

158

160

161

162

74

Pro- and mesonotum each with three keels, lateral keels of pronotum caudally curved outwards. Antennae short. Teeth of post-tibial calcar with tendency of reduction. Frons below with two black or fuscous spots, clypeus light without a tendency of darkening. Wing dimorphous. Genital scale of female reniform, caudal margin thickened, sclerotic. In Denmark and Fennoscandia two species.

Key to species of *Stiroma*

1 Styles of male apically simple (Text-fig. 165). Median border of lateral lobes of female straight in major part of its length (Text-fig. 168). Genital scale smaller, width less than 0.2 mm (Text-fig. 169)   23. *bicarinata* (Herrich-Schäffer)
– Styles of male apically ending in a two-pointed dilatation (Text-fig. 172). Lateral lobes of female broadest in their caudal 1/3, their median border faintly S-curved. Genital scale of female larger, width 0.2 mm or more (Text-fig. 176)

24. *affinis* Fieber

23. *Stiroma bicarinata* (Herrich-Schäffer, 1835)
    Plate-fig. 4, text-figs. 163–169.

Delphax bicarinata Herrich-Schäffer, 1835: 66.
Delphax mutabilis Boheman, 1847a: 43.
Delphax nasalis Boheman, 1847a: 41.

Brownish yellow or sordid light yellow. Black spots on frons as stated in our diagnosis of the genus, in addition to these are some usually well-marked black spots on genae and thorax. Mesonotum in brachypters brownish yellow with black lateral corners. Abdomen of an uniform light colour or on each side with a more or less well-marked blackish longitudinal band. In dark specimens the abdomen may be of an uniform blackish colour, and the black spots on anterior part of body may be more extended. In macropterous dark specimens, for instance mesonotum may be uniformly fuscous. In the brachypterous form (Plate-fig. 4), which is normally much more common than the macropterous form, the fore wings ar semi-hyaline, caudally either rounded or fairly squarely cut, rarely reaching beyond the hind border of the 3rd abdominal segment. The wings of macropters reach considerably behind apex of abdomen, and are hyaline with somewhat darker veins. Pygofer of male as in Text-fig. 163, anal segment of male as in Text-fig. 164, style as in Text-fig. 165, aedeagus as in Text-figs. 166 and 167. Venter

Text-figs. 155–162. *Eurysula lurida* (Fieber). – 155: male pygofer from behind; 156: male pygofer from the right; 157: male anal tube from behind; 158: genital style; 159: aedeagus, ventral aspect; 160: aedeagus from the left; 161: aedeagus from the right; 162: caudal part of female abdomen from below. Scale: 0.25 mm for 162, 0.1 mm for the rest.

of posterior part of female abdomen, see Text-fig. 168, female genital scale as Text-fig. 169. Overall length of brachypters 3–4 mm, macropters 4.5–5 mm.

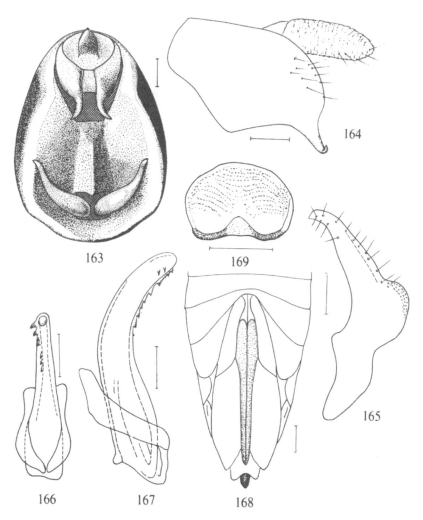

Text-figs. 163–169. *Stiroma bicarinata* (Herrich-Schäffer). – 163: male pygofer from behind; 164: male anal tube form the left; 165: left genital style from outside; 166: aedeagus in ventral aspect; 167: aedeagus from the left; 168: caudal part of female abdomen from below; 169: genital scale of female from below. Scale: 0.25 mm for 168, 0.1 mm for the rest.

Distribution: Fairly common in Denmark (SJ, EJ, F, LFM, SZ, NWZ, NEZ, B). – Common in Sweden, Sk. – Ås. Lpm. – In Norway so far established only in Os, Bø, and Nsi. – Common in East Fennoscandia, found up to LkW and Lr. – Widespread in Europe, also found in Tunisia, Kazakhstan, W. Siberia and Mongolia.

Biology. In woods (Kuntze, 1937). In low vegetation in light leafy wood (Wagner & Franz, 1961). Also "in older leys, from where it rather frequently migrated to cereal crops. – The flight period was 24.6.–12.7. – The species was a daytime flier and flight was observed when the daily maximum was at least 18°C." (Raatikainen & Vasarainen, 1973). Adults in June–August.

24. *Stiroma affinis* Fieber, 1866
    Text-figs. 170–176.

Stiroma affinis Fieber, 1866b: 531.

Very like the preceding species from which it can be separated only by the characters given in the key. Male pygofer as in Text-fig. 170, anal tube of male as in Text-fig. 171, style as in Text-fig. 172, aedeagus as in Text-figs. 173, 174. Ventral aspect of posterior part of female abdomen as in Text-fig. 175, genital scale as in Text-fig. 176. Overall length of brachypters 2.5–3 mm, of macropters about 5 mm.

Distribution. Widespread and fairly common in Denmark. – Fairly common also in Sweden, found in most provinces up to Lu. Lpm. – In Norway so far established in Os, Bø, TEi, and HOi. – In East Fennoscandia less common than *bicarinata,* but common in the south-west, found up to Li and Lr. – Widespread in Europe, also found in Kazachstan, Altai, and Mongolia.

Biology. In woods and glades (Kuntze, 1937). In low vegetation in light leafy woods (Wagner & Franz, 1961). Hibernates in larval stages (Müller, 1957).

Genus *Stiromoides* Vilbaste, 1971

Stiromoides Vilbaste, 1971: 133.

Type-species: *Eurysa maculiceps* Horváth, 1903, by original designation.

As *Eurysa,* head comparatively longer, anteriorly with a large shining black spot. Frons with two indistinct median carinae. One species.

25. *Stiromoides maculiceps* (Horváth, 1903)
    Text-figs. 177–184.

Eurysa maculiceps Horváth, 1903: 475.
Eurysa maculiceps Linnavuori, 1952b: 73.
Stiromoides maculiceps Vilbaste, 1971: 134.

77

Fore body whitish – ochraceous yellow. Frons comparatively narrow, its length being about 1.35 × its width, with more or less evenly arched sides, its greatest width about the level of lower margins of eyes. Frons below with a transverse black stripe which is joined with a large shining black spot occupying central part of vertex and upper part

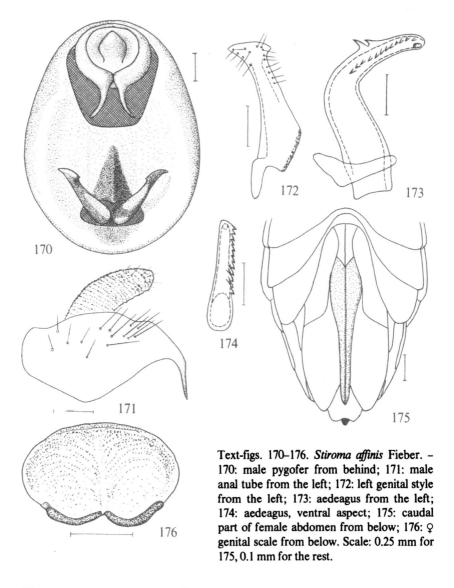

Text-figs. 170–176. *Stiroma affinis* Fieber. – 170: male pygofer from behind; 171: male anal tube from the left; 172: left genital style from the left; 173: aedeagus from the left; 174: aedeagus, ventral aspect; 175: caudal part of female abdomen from below; 176: ♀ genital scale from below. Scale: 0.25 mm for 175, 0.1 mm for the rest.

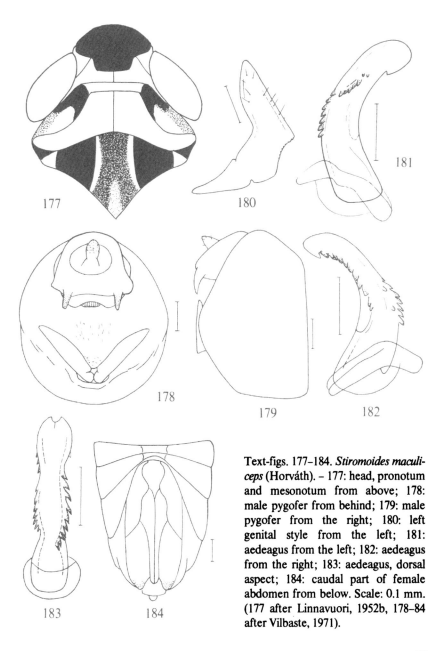

Text-figs. 177–184. *Stiromoides maculiceps* (Horváth). – 177: head, pronotum and mesonotum from above; 178: male pygofer from behind; 179: male pygofer from the right; 180: left genital style from the left; 181: aedeagus from the left; 182: aedeagus from the right; 183: aedeagus, dorsal aspect; 184: caudal part of female abdomen from below. Scale: 0.1 mm. (177 after Linnavuori, 1952b, 178–84 after Vilbaste, 1971).

of frons by a black median stripe. Clypeus light. Edges of pronotum anteriorly blackish. Mesonotum blackish – fuscous with two narrow longitudinal stripes and two lateral spots yellowish (Text-fig. 177). Fore wings of brachypters little shorter than abdomen, dull, semi-transparent. Abdomen of male above yellowish, laterally blackish. Venter, legs and antennae dirty brown-yellow. Genital segment of male (Text-figs. 178, 179) yellowish, laterally with a black spot, anal tube large, underneath blackish, with two short appendages. Styles (Text-fig. 180) comparatively long, dark brown. Aedeagus as in Text-figs. 181–183. Abdomen of female brownish, hind borders of terga medially narrowly, laterally more broadly, light. Lateral sides of female abdomen dark-shaded with blackish spots on hind margins of posterior segments and of pygofer. Pygofer concolorous, saw-case a little darker. Abdomen below as in Text-fig. 184. Total length of brachypters 2–2.4 ($\male$) to 3.3–3.5 ($\female$) mm.

Distribution. In our countries found only in Finland, Sa: Joutseno (Linnavuori). – Present in Hungary, Estonia, Kazakhstan, Siberia, and Mongolia.

Biology. On dry meadows, adults in July and August.

Note. I have not seen this species. The description has been compiled and the illustrations copied from papers by Linnavuori (1952b) and Vilbaste (1971).

SUBFAMILY ACHOROTILINAE

First tarsal segment and aedeagus as in Stirominae. Carinae of frons strong, median carina simple, forked or double. Frons often narrower above than below, without a tendency of vaulting. Vertex not narrowing. Frons with a tendency of developing whitish spots. Basal antennal segments not enlarged. Teeth of post-tibial calcar weakly developed.

Genus *Achorotile* Fieber, 1866

Achorotile Fieber, 1866b: 521.
Type-species: *Delphax albosignata* Dahlbom, 1850, by monotypy.

Frons with two sharp and distinct median carinae somewhat converging downwards. Vertex approximately as long as broad, with parallel sides, anteriorly reaching a little in front of compound eyes. Pronotum caudally obtuse-angled concave, with three carinae; lateral carinae curved towards sides. Mesonotum with three carinae, the median one weak. Frons laterally of median keels, pronotum along lateral keels, and abdomen with a number of sensory pits (as in larval instars of all Delphacidae). In Fennoscandia two species.

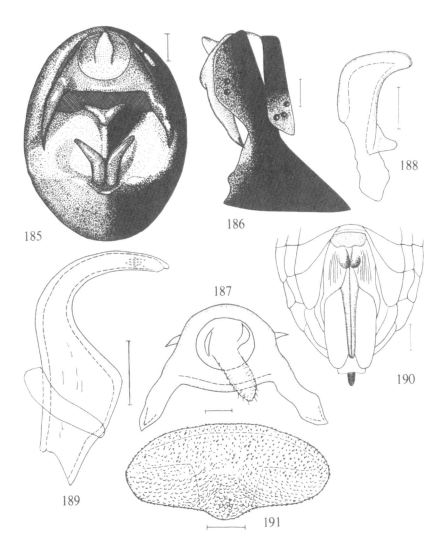

Text-figs. 185–191. *Achorotile albosignata* (Dahlbom). – 185: male pygofer from behind; 186: male pygofer from the right; 187: male anal tube from behind; 188: right genital style obliquely from outside; 189: aedeagus from the left; 190: caudal part of female abdomen from below; 191: genital scale of female from below. Scale: 0.25 mm for 190, 0.1 mm for the rest.

Key to species of *Achorotile*

1 First antennal segment almost twice as long as broad. Second antennal segment by 1/3 longer than first segment, and twice as long as broad 26. *albosignata* (Dahlbom)
– First antennal segment 2.5 times as long as basal width. Second antennal segment 1.5 times as long as first segment, and more than 3 times as long as broad
27. *longicornis* (J. Sahlberg)

26. ***Achorotile albosignata*** (Dahlbom, 1850)
Plate-fig. 8, text-figs. 185–191.

Delphax albosignata Dahlbom, 1850: 199.
Delphax fuscinervis Boheman, 1852b: 113.
Achorotile albosignata Fieber, 1866b: 521.

Wing dimorphous. The brachypterous male is black with a broad, ivory-white, longitudinal stripe on vertex, pronotum and scutellum; fore wings brownish. Median 2/3 of third and fourth abdominal terga more or less broadly ivory-white along their caudal margins, and a narrow median longitudinal stripe on the 5th–7th abdominal terga is also ivory-white. Underside of body and legs largely fuscous. Fore wings of brachypters apically squarely truncate reaching just beyond hind border of 2nd abdominal tergum. The brachypterous female (Plate-fig. 8) may look more or less like the male in colour, but often she is considerably lighter with frons yellow-brownish, especially above, yellowish fore wings and sordid yellow legs. Fore wings of the macropterous form whitish hyaline with dark veins. Length of brachypters 2.25–3.25 mm, of macropters 3.5 mm. Male pygofer as in Text-figs. 185 and 186, anal tube of male as in Text-fig. 187, styles as in Text-fig. 188, aedeagus as in Text-fig. 189. Venter of posterior part of female abdomen as in Text-fig. 190, lateral lobes each with a basal protuberance. Genital scale large (Text-fig. 191).

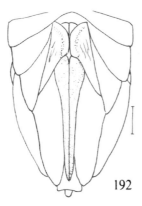

192

193

Text-figs. 192, 193. *Achorotile longicornis* (J. Sahlberg). – 192: caudal part of female abdomen from below; 193: ♀ genital scale from below. Scale: 0.25 mm for 192, 0.1 mm for 193.

Distribution. Not yet found in Denmark. – Not uncommon in Sweden, Sk. – T. Lpm. – In Norway only found by Holgersen in HEn: Folldals Verk, and TEi: Møsvatn. – Scarce and sporadic in East Fennoscandia, found in Al, Ab, N, St, Ta, Tb, Sb, Ks, Le, Li, Kr, and Lr. – Found in France, DDR, n. Italy, n. Poland, n. and m. Russia, w. Siberia.

Biology. On *Agrostis canina* and other grasses on rocks (Linnavuori, 1969). Adults in June–August.

27. *Achorotile longicornis* (J. Sahlberg, 1871)
    Text-figs. 192, 193.

Ditropis longicornis J. Sahlberg, 1871: 474.

Closely resembling *albosignata*. Specimens so far seen by me are more brownish (by fading?). Antennae longer, vertex and frons narrower, and there is an ivory-white transverse stripe also on 5th abdominal tergum. I have only seen brachypterous females of this species. Venter of posterior part of female abdomen as in Text-fig. 192. Genital scale (Text-fig. 193) much smaller than in *albosignata*. Length of brachypterous female about 3 mm.

Distribution. Not found in Denmark, nor in Norway. – Very rare in Sweden and East Fennoscandia. Anton Jansson found one specimen in Messaure 17.VII.1952, and another in Pålkem 18.VII.1952, both localities in Lu. Lpm. Palmén found *longicornis* in Finland, N: Helsinge; J. Sahlberg collected it in Sa: Parikkala, and in Kirvajalahti, Petrosavoksk, Tiudie, and Juustjärvi in Kr. – Poland (?), n. Russia.

Biology. Adults in July–August, on dry sandy zootopes.

Genus *Euconomelus* Haupt, 1929

Euconomelus Haupt, 1929: 212.
Type-species: *Delphax lepida* Boheman, 1847, by original designation.

Frons almost hexagonal. Median carina of frons forking somewhat below junction with vertex. Veins of fore wings with conspicuous dark tubercles. Antennae cylindrical without keels. One species.

28. *Euconomelus lepidus* (Boheman, 1847)
    Plate-figs. 9, 28, text-figs. 194–202.

Fulgora limbata Fabricius, 1794: 6 (non Olivier, 1791: 574).
Delphax lepida Boheman, 1847b: 265.
Delphax tristis Boheman, 1847a: 60.

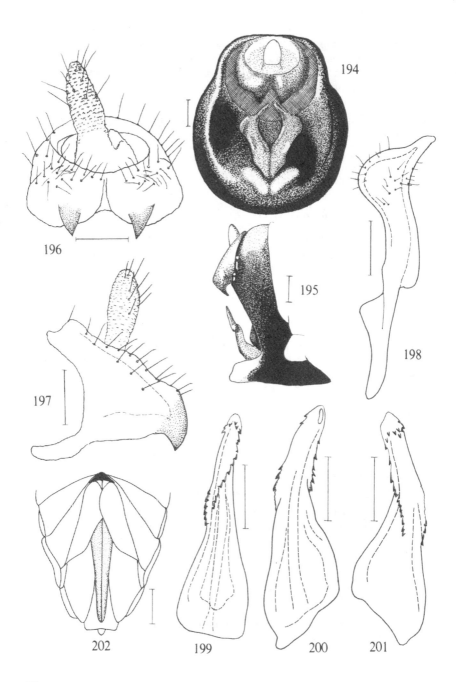

194

195

196

197

198

199

200

201

202

84

Body yellowish brown to blackish brown. Frons dark brown with several short, yellowish white, transverse stripes. Tibiae yellowish brown, each with two brown rings, hind tarsi with one brown ring. Proximal half of fore wings of brachypters yellowish white, except for median angle which is brownish; apical half dark brown with 3 whitish spots on apical border. Especially in the male abbreviated fore wings are obliquely cut off. Fore wings of macropters (Plate-fig. 9) whitish or colourless hyaline with inner proximal corner, an oblique transverse band on proximal half, another band often divided into spots on transverse veins, and spots on apices of longitudinal veins dark brown or blackish brown. Abdomen dark brown or blackish brown with several longitudinal rows of light spots. Male pygofer as in Text-figs. 194 and 195, anal tube as in Text-figs. 196 and 197, style as in Text-fig. 198, aedeagus as in Text-figs. 199–201. Ventral aspect of posterior part of female abdomen, see Text-fig. 202. Lateral lobes of female overlapping one another, genital scale large, black. Length of brachypters 1.5–2.7 mm, length of macropters with wings 3.0–3.9 mm.

Distribution. Scarce in Denmark but found in all provinces except LFM, SZ, and NWZ. – Not rare in the south of Sweden, found in Sk., Bl., Hall., Öl., Gtl., Ög., Sdm., Upl., and Vstm. – Norway: found in Bø, VAy, and Ry. – Common in Southern and Central Finland, found in Al, Ab, N, St, Sa, Ok, and ObS. – Widespread in Europe, present also in Altai, Kazakhstan, Uzbekistan, w. and m. Siberia, and Mongolia.

Biology. Locally abundant on *Cyperaceae* in wet zootopes, adults in July–September. Macropters rare.

SUBFAMILY DELPHACINAE

1st tarsal segment of hind legs with 5 + 2 or 6 + 2 spines. Vertex, frons, and carinae of frons as in Achorotilinae. Basal antennal segments more or less enlarged. Teeth of post-tibial calcar well developed. Aedeagus without traces of theca and membranous apical part.

Genus *Conomelus* Fieber, 1866

Conomelus Fieber, 1866b: 520.
Type-species: *Delphax anceps* Germar, 1821, by designation under the Plenary Power of the Int. Commission.

Text-figs. 194–202. *Euconomelus lepidus* (Boheman). – 194: male pygofer from behind; 195: male pygofer from the right; 196: male anal tube from behind; 198: genital style; 199: aedeagus, dorsal aspect; 200: aedeagus from the right; 201: aedeagus from the left; 202: caudal part of female abdomen from below. Scale: 0.25 mm for 202, 0.1 mm for the rest.

Frons almost hexangular. Median carina of frons forking somewhat below junction with vertex. First antennal segment flattened, with one longitudinal keel on the front and another on the back (Text-fig. 203). Second antennal segment with a short basal keel on the back. Fore wings with conspicuous dark tubercles on the veins. In North Europe one species.

29. Conomelus anceps (Germar, 1821)
Plate-fig. 29, text-figs. 203–210.

Delphax anceps Germar, 1821: 105.
Delphax signifera Boheman, 1845b: 164.
Delphax palliata Boheman, 1847b: 266.
Conomelus limbatus Fieber, 1866b: 520, et auct. (non Fabricius, 1794).

Brownish yellow, lower frons and clypeus darker. Fore wings of brachypters apically truncate, hyaline with a dark spot in each apical angle, or these spots may be joined together into a transverse band. Fore wings of macropters with various dark brown markings, among these a longitudinal streak in clavus near its apex along the commissural border, and a sickle-shaped, often ramified, band in the apical part of the wing. Abdomen dark mottled or dark brown with longitudinal rows of light spots, or almost wholly dark brown. Male pygofer as in Text-figs. 204 and 205, style as in Text-fig. 206, aedeagus as in Text-figs. 207–209. Venter of female abdomen as in Text-fig. 210. Overall length of brachypters 2.1–3.5 mm, of macropters 4.0–4.15 mm.

Distribution. Common in Denmark, found in all districts except SJ and SZ. – In Sweden common up to Hls. – In Norway up to HOy and HEs, especially in the coastal parts. – Common in Southern and Central East Fennoscandia, found in Al, Ab, N, Ka, St, Ta, Sa, Kb, Vib. – Widespread in Europe, also found in Algeria.

Biology. Often abundant on *Juncus* spp. in wet biotopes. Adults in June–September. Hibernation in the egg stage, according to Müller (1957). The macropterous form is rather rare.

Genus *Delphax* Fabricius, 1798

Delphax Fabricius, 1798: 511.
Type-species: *Cicada crassicornis* Panzer, 1796, by subsequent designation under the Plenary Powers of the International Commission.

Araeopus Spinola, 1839: 336.
Type-species: *Cicada crassicornis* Panzer, 1796, by monotypy.

Frons with a median carina. Vertex broad, anteriorly almost straight, reaching just in front of eyes. Antennae long, their first segment longer than the second one, flattened and with keels. Pronotum about as long as vertex, its fore border behind vertex straight, its hind border faintly curved, lateral carinae curving outwards. Mesonotum with three carinae. Wing-polymorphous species with a macropterous, a brachypterous and an intermediary form. In the latter the wings are distinctly longer than abdomen but shorter

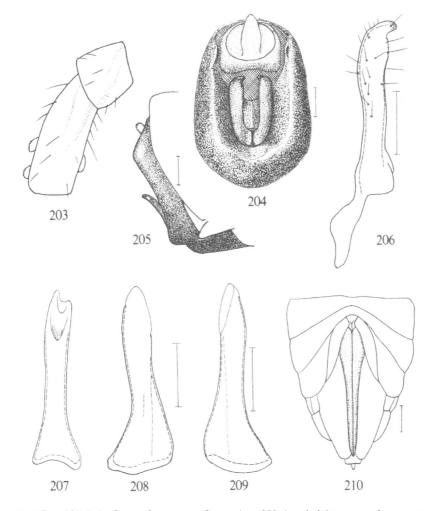

Text-figs. 203–210. *Conomelus anceps* (Germar). – 203: basal right antennal segments from inside; 204: male pygofer from behind; 205: male pygofer from the right; 206: genital style; 207: aedeagus, dorsal aspect; 208: aedeagus from the right; 209: aedeagus from the left; 210: caudal part of female abdomen from below. Scale: 0.25 mm for 210, 0.1 mm for the rest.

than those of macropters. In Denmark and Fennoscandia two species, our largest of the family *Delphacidae*, both living on reed *(Phragmites communis)*.

Key to species of *Delphax*

1 Fore wing of macropters (♂♀) with a strongly marked, zigzag-shaped, comparatively broad, on middle sometimes interrupted, black-brown longitudinal band with a narrower branch towards costal margin arising in the apical part of the wing. Fore wing of the brachypterous female with an oblique black-brown longitudinal streak from wing basis directed towards apical margin. Appendages of anal tube of male unsymmetrical (Text-figs. 213, 214). Visible part of genital scale in female semicircular (see Text-fig. 219)   30. *crassicornis* (Panzer)

– Fore wing of macropters in both sexes with a straight longitudinal band just outside claval suture reaching ± to level of apex of clavus, and in apical part of wing with a sickle-shaped band without a distinct connection with the former one. Appendages of anal tube in male symmetrical (Text-fig. 222). Visible part of genital scale in female oblong, approximately parallel-sided (see Text-fig. 228). Fore wings of brachypterous female without dark markings   31. *pulchellus* (Curtis)

30. *Delphax crassicornis* (Panzer, 1796)
Plate-figs. 10, 11, text-figs. 211–219.

Cicada crassicornis Panzer, 1796: 19.
Cicada dubia Panzer, 1796: 20.
Delphax crassicornis Fallén, 1826: 73.

Yellowish brown. Frons above clypeus with a whitish transverse band and on middle with a narrower transverse band continuing to and behind attachment of antennae. Pronotum laterally with a broad, blackish brown, longitudinal band continuing past side corners of mesonotum and – though usually less distinctly – on basis of fore wing. Macropterous and intermediary males and macropterous females have transparent fore wings with dark markings (see key to species and Plate-fig. 10). Most females are brachypterous with apically rounded fore wings covering basal 1/2-2/3 of abdomen (Plate-fig. 11). Legs with dark streaks and spots. Abdomen of male in its major part black. Pygofer of male as in Text-figs. 211, 212, anal tube of male as in Text-figs. 213, 214, style as Text-fig. 215, aedeagus as in Text-figs. 216–218. Venter of caudal part of female abdomen as in Text-fig. 219. Length of brachypterous female 4.8–5.4 mm, of macropterous male (with wings) 5.1–6 mm, of macropterous female 7 mm.

Distribution. Scarce in Denmark, found in EJ, WJ, NWJ, NEJ, LFM, NEZ, and B. – Scarce also in Sweden, Sk. – Dlr. – So far not found in Norway. – Rare and sporadic in East Fennoscandia, recorded from Ab, N, Kb, and Kr. – Not in Great Britain, nor in

France, widespread in remaining part of Europe, also in Kazakhstan, Kirghizia, Tadzjikistan, and Japan.

Biology. On *Phragmites communis*, sometimes abundant (Kontkanen, 1938), adults in July–August.

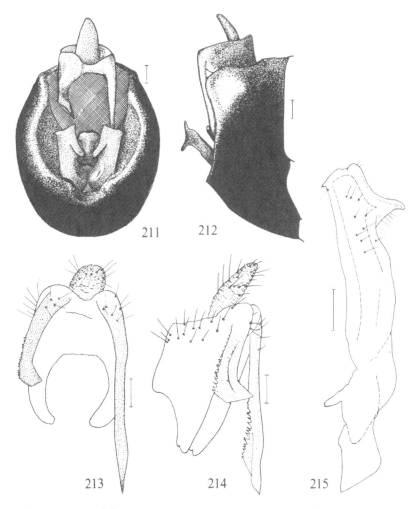

Text-figs 211–215. *Delphax crassicornis* (Panzer). – 211: male pygofer from behind; 212: male pygofer from the right; 213: male anal tube from behind; 214: male anal tube from the left; 215: left genital style. Scale: 0.1 mm.

89

31. *Delphax pulchellus* (Curtis, 1833)
Plate-fig. 12, text-figs. 220–228.

Asiraca pulchella Curtis, 1833, Pl. 445.
Areopus (sic) *minki* Fieber, 1866b: 522.
Delphax minki J. Sahlberg 1871: 401.

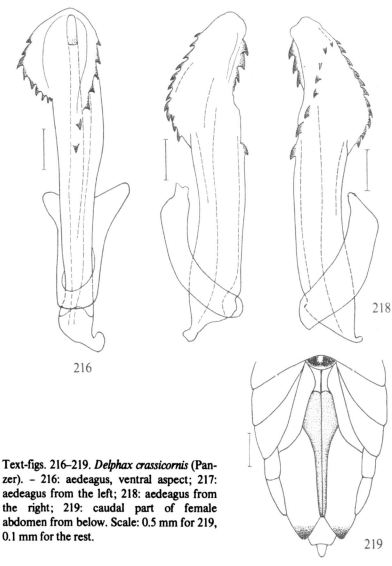

Text-figs. 216–219. *Delphax crassicornis* (Panzer). – 216: aedeagus, ventral aspect; 217: aedeagus from the left; 218: aedeagus from the right; 219: caudal part of female abdomen from below. Scale: 0.5 mm for 219, 0.1 mm for the rest.

Resembling *crassicornis* but dark markings less extended, paler and less well-marked. Fore wing of macropters, see Plate-fig. 12. Male pygofer as in Text-figs. 220 and 221, anal segment of male as in Text-figs. 222, 223, style as in Text-fig. 224, aedeagus as in Text-figs. 225–227. Venter of posterior part of female abdomen as in Text-fig. 228. Overall length of brachypterous female 5 mm, of intermediary male 5.5–6 mm, of macropterous male 6.5 mm, of macropterous female 7–7.5 mm.

Distribution. Fairly common in Denmark (SJ, NWJ, NEJ, F, LFM, SZ, B). – Rare in Sweden: Sk., Svalöf (Ossiannilsson), Bl. (Boheman), Gtl., Fårön, Alnös Träsk 14.VIII.-

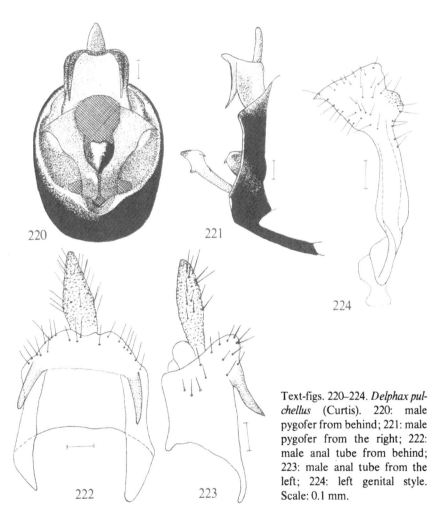

220

221

224

222

223

Text-figs. 220–224. *Delphax pulchellus* (Curtis). 220: male pygofer from behind; 221: male pygofer from the right; 222: male anal tube from behind; 223: male anal tube from the left; 224: left genital style. Scale: 0.1 mm.

91

1931 (Lohmander), Sdm., Dalarö (Reuter). – Norway: only found in VAy: Kristiansand 21.VII.–9.VIII, in 1929–36, by Warloe. – In East Fennoscandia commoner than *crassicornis*, found in Al, Ab, N, and St. – Widespread in Europe.

Biology. On *Phragmites communis* especially at the seaside, adults in July–August.

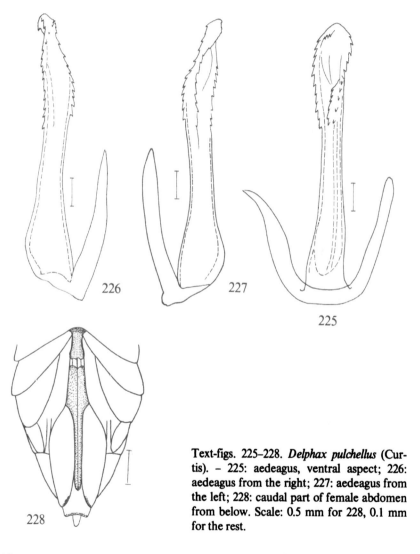

Text-figs. 225–228. *Delphax pulchellus* (Curtis). – 225: aedeagus, ventral aspect; 226: aedeagus from the right; 227: aedeagus from the left; 228: caudal part of female abdomen from below. Scale: 0.5 mm for 228, 0.1 mm for the rest.

Genus *Euides* Fieber, 1866

Euides Fieber, 1866b: 519.
Type-species: *Delphax basilinea* Germar, 1821, by subsequent designation.
Euidella Puton, 1886: 72.
Type-species: *Delphax basilinea* Germar, 1821, by subsequent designation.

Vertex almost square, reaching in front of eyes by 2/5 of its length. Side borders of frons below eyes almost parallel. Frons somewhat narrower between eyes, its median carina sharp, simple, forking just below vertex. First and second antennal segments fairly long, not flattened. Pro- and mesonotum each with 3 distinct carinae. In North Europe one species.

32. *Euides speciosa* (Boheman, 1845)
Plate-fig. 13, text-figs. 229–235.

Delphax speciosa Boheman, 1845b: 165.
Euides speciosa Fieber, 1866b: 519, 532.

In colour much resembling *Delphax* spp., but easily separated from these e. g. by structure of head and antennae. Wing dimorphous. The male (Plate-fig. 13) is always macropterous, brownish yellow with a broad white longitudinal band on vertex, pronotum and mesonotum, and with abdomen black. Fore wings semi-transparent, whitish with the following blackish brown markings: an oblong triangular spot in corium from wing basis along basal half of corioclaval suture, a small narrow streak in apex of clavus, and a sickle-shaped patch in the apical part of the wing. The macropterous female resembles the male, but its abdomen is mottled in lighter and darker brown, and the markings of the fore wings are lighter and more diffuse. The brachypterous female is entirely brownish yellow with transparent apically rounded fore wings covering about half abdomen, or with more or less extended blackish brown markings especially on abdomen. In such dark specimens the fore wings may also show traces of the markings of macropters in the shape of dark longitudinal streaks in the basal cells and dark spots on the apical border. Male pygofer as in Text-figs. 229 and 230, style as in Text-fig. 231, aedeagus as in Text-figs. 232–234. Venter of caudal part of female abdomen as in Text-fig. 235. Overall length of male 5–5.8 mm, brachypterous female 4.5–5 mm, macropterous female 6.5–6.75 mm.

Distribution. Scarce in Denmark, found in SJ, EJ, NWJ, NEJ, F, NEZ, and B. – Not rare in Sweden (Sk., Bl., Vg., Sdm., Upl., and Nrk.). – Not recorded from Norway. – Very rare in Finland, found in Al, Ab, Oa, and Kb. Also in Kr. – Widespread in Europe, also found in Kazakhstan.

Biology. On *Phragmites communis*. Halophilous, according to Linnavuori (1952a). Adults in June–August.

93

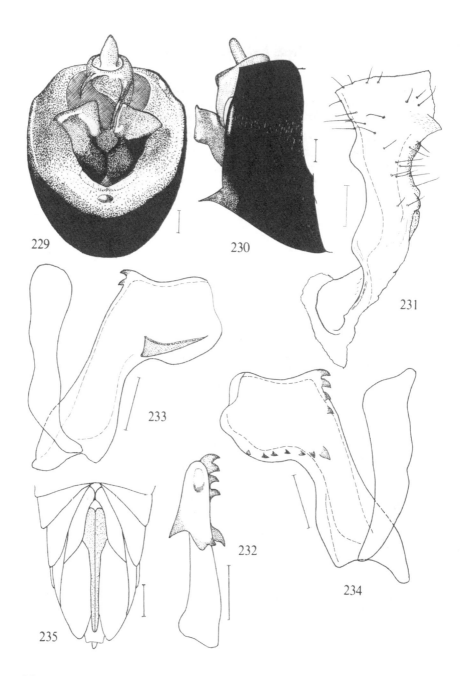

229

230

231

233

232

234

235

SUBFAMILY CHLORIONINAE

Vertex more or less narrow, posteriorly broader. Frons of last-instar larvae broadly oval. Frons of adults broadened at middle or below middle. No tendency of developing whitish spots on frons. Body often greenish. Males often with tooth-like processes on upper side of anal segment.

Genus *Chloriona* Fieber, 1866

Chloriona Fieber, 1866b: 519.
  Type-species: *Delphax unicolor* Herrich-Schäffer, 1835, by subsequent designation.

Frons (Text-figs. 237, 274) widest below middle, with a simple, sharp median carina. Vertex (Text-fig. 236) narrowed in front, reaching considerably in front of eyes. 1st antennal segment twice as long as broad, shorter than second segment. Hind margin of pronotum obtuse-angled concave, lateral keels of pronotum suddenly curving outwards or obsolescent before reaching hind margin (Text-fig. 236). Post-tibial calcar long, with numerous (up to 30) marginal teeth. Wing-polymorphous species all living on reeds *(Phragmites)* and much alike. Separation of females is especially difficult. In Fennoscandia and Denmark five species.

Key to species of *Chloriona*

1	Males	2
–	Females	6
2 (1)	Caudal border of pygofer approximately as high as broad (Text-figs. 238, 257). Styles apically broadened, each apically with two acute angles (Text-figs. 241, 260)	3
–	Pygofer as seen from behind broader than high	4
3 (2)	Pygofer black (Text-fig. 238). Styles sharply narrowing near apex (Text-fig. 241)	33. *smaragdula* (Stål)
–	Pygofer largely pale (Text-fig. 257). Styles gradually narrowed towards apex (Text-fig. 260)	35. *dorsata* Edwards
4 (2)	Pygofer entirely or largely black. Styles strongly diverging, S-curved (Text-figs. 248, 251)	34. *chinai* Ossiannilsson
–	Pygofer largely whitish-yellow	5

Text-figs. 229–235. *Euides speciosa* (Boheman). – 229: male pygofer from behind; 230: male pygofer from the right; 231: left genital style; 232: aedeagus, ventral aspect; 233: aedeagus from the left; 234: aedeagus from the right; 235: caudal part of female abdomen from below. Scale: 0.5 mm for 235, 0.1 mm for the rest.

5 (4) Styles apically indistinctly thickening (Text-fig. 268). Upper teeth of anal tube widely apart (Text-fig. 267) 36. *glaucescens* Fieber
– Styles apically obliquely truncate with an acute angle directed outwards – downwards (Text-fig. 277). Upper teeth of anal tube close (Text-fig. 276)
37. *vasconica* Ribaut
6 (1) Lateral lobes proximally each with an angular projection, which entirely or almost entirely conceals basis of ovipositor (Text-fig. 272) 36. *glaucescens* Fieber
– Proximal projection of lateral lobes, if present, not angular 7
7 (6) Lateral lobes proximally virtually without any projection (Text-fig. 264)
35. *dorsata* Edwards
– Lateral lobes each with a proximal rounded medially directed projection 8
8 (7) Saw of ovipositor with fine and dense teeth (Text-fig. 256) 34. *chinai* Ossiannilsson
– Saw of ovipositor with coarser and less densely arranged teeth (Text-figs. 247, 282)
9
9 (8) Frons widest considerably below eyes, sides often almost angular (Text-fig. 237)
33. *smaragdula* (Stål)
– Frons widest near lower margin of eyes or just below eyes, sides as a rule smoothly rounded (Text-fig. 274) 37. *vasconica* Ribaut

33. *Chloriona smaragdula* (Stål, 1853)
Plate-fig. 14, text-figs. 236–247.

Delphax smaragdula Stål, 1853: 174.
Chloriona smaragdula J. Sahlberg, 1871: 407.
Chloriona prasinula Fieber, 1872: 5.
Chloriona prasinula Jensen-Haarup, 1920: 37.

Male (Plate-fig. 14) macropterous, female sometimes also macropterous but usually brachypterous with fore wings about one and a third times as long as broad, apically rounded, leaving major part of abdomen uncovered. Fore part of body in male whitish-yellow, scutellum sometimes with dark patch laterally of side keels, abdomen in its greater part with side and segment margins yellowish or reddish yellow. Fore wings twice as long as abdomen, semi-hyaline, whitish, with veins concolorous and provided with short fine black setae. The brachypterous female is light green with yellowish segment borders and yellowish fore wings. The green colour of the body is delicate, wherefore dead specimens are usually yellow, not green. Apices of pygofer often darkened. Especially in females from northern localities black pigmentation may be more or less extended on the entire abdomen. The macropterous female resembles the male in colour but its abdomen is largely yellowish. Male pygofer as in Text-fig. 238, male anal tube as in Text-figs. 239, 240, style as in Text-fig. 241, aedeagus as in Text-figs. 242–245, venter of apical part of female abdomen as in Text-fig. 246, saw of ovipositor as in Text-fig. 247, comparatively coarsely serrate. Length of macropters 4–6 mm, of brachypters 4.75–5.6 mm.

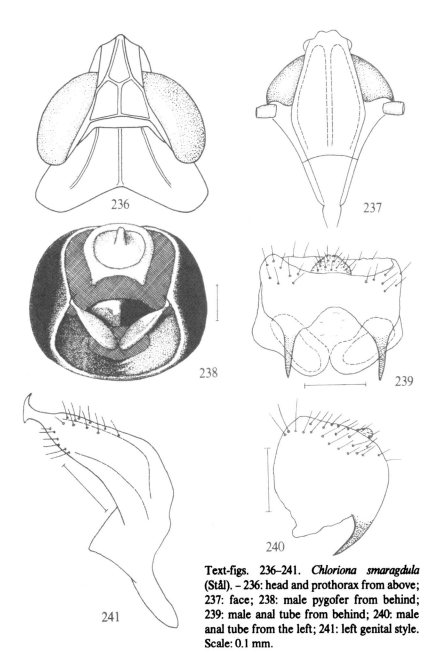

Text-figs. 236–241. *Chloriona smaragdula* (Stål). – 236: head and prothorax from above; 237: face; 238: male pygofer from behind; 239: male anal tube from behind; 240: male anal tube from the left; 241: left genital style. Scale: 0.1 mm.

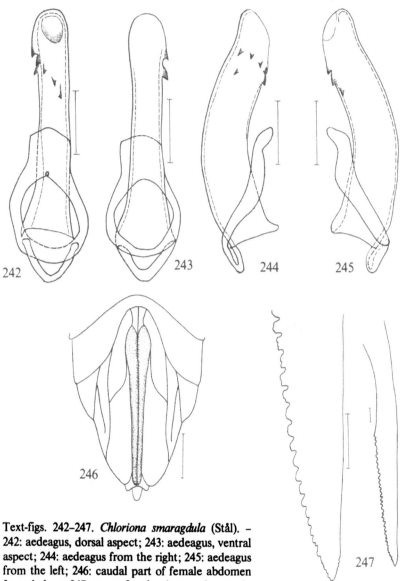

Text-figs. 242–247. *Chloriona smaragdula* (Stål). –
242: aedeagus, dorsal aspect; 243: aedeagus, ventral
aspect; 244: aedeagus from the right; 245: aedeagus
from the left; 246: caudal part of female abdomen
from below; 247: saw of ovipositor and apex in
higher magnification. Scale: 0.5 mm for 246, 0.1 mm
for the rest.

Distribution. Common in Denmark, found in all Danish provinces except SJ and WJ. – Common in Sweden, found from Sk. up to Lu. Lpm. – Not yet recorded from Norway. – Scarce in East Fennoscandia, found in Al, N, Oa, Kb, Om, and Kr. – Widespread in Europe, established also in Kazakhstan and w. Siberia.

Biology. On *Phragmites,* adults June–August. Hibernates in larval stages according to observations of Strübing in Germany (Müller, 1957).

34. *Chloriona chinai* Ossiannilsson, 1946
Text-figs. 248–256.

Chloriona prasinula Lindberg, 1935: 115 (nec Fieber, 1872).
Chloriona ? prasinula Ossiannilsson, 1943: 18 (nec Fieber).
Chloriona chinai Ossiannilsson 1946b: 86.

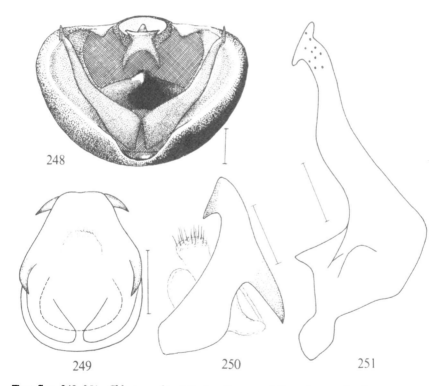

Text-figs. 248–251. *Chloriona chinai* Ossiannilsson. – 248: male pygofer from behind; 249: male anal tube from behind; 250: male anal tube from the left; 251: left genital style. Scale: 0.1 mm.

In colour very like *smaragdula*. The male sex is easily separated by the different structure of the genitalia. The female differs from our remaining *Chloriona* species by the finer serration of the saw of the ovipositor. Male pygofer from behind as in Text-fig. 248, anal tube of male as in Text-figs. 249, 250, style as in Text-fig. 251, aedeagus as in Text-figs. 251–254. Ventral aspect of posterior part of female abdomen as in Text-fig. 255, saw of female ovipositor as in Text-fig. 256. Length of macropters 4.2–5.7 mm, of brachypters 4.6–5.1 mm.

Distribution. So far not found in Denmark, nor in Norway. – Comparatively common in Sweden, found in Sk., Ög., Vg., Upl., and Nb. – Scarce in East Fennoscandia, established in Al, Ab, N, Ta, Oa, Kb, Om, and Kr. – Also found in Estonia, north Russia, and Siberia.

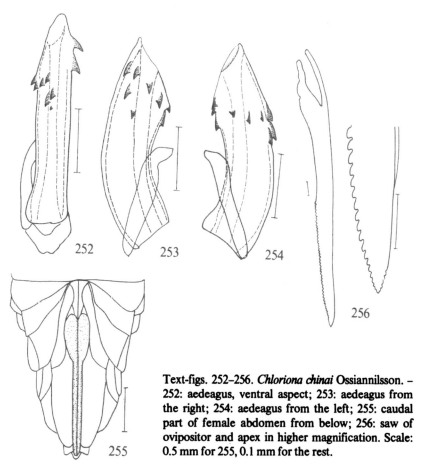

Text-figs. 252–256. *Chloriona chinai* Ossiannilsson. – 252: aedeagus, ventral aspect; 253: aedeagus from the right; 254: aedeagus from the left; 255: caudal part of female abdomen from below; 256: saw of ovipositor and apex in higher magnification. Scale: 0.5 mm for 255, 0.1 mm for the rest.

Biology. On *Phragmites*, adults in June–August.

35. *Chloriona dorsata* Edwards, 1898.
Text-figs. 257–265.

Chloriona dorsata Edwards, 1898: 59.
Chloriona danica Jensen-Haarup, 1917: 4.

Very like *Chloriona smaragdula*, males differing by details in their genitalia, females by the shape of their lateral lobes. In the literature the pigmentation of the ninth abdominal sternum is being used as a character for the separation of females of *dorsata* from certain other *Chloriona* species, but I do not think that this is a reliable character since there is a considerable variation in pigmentation in *smaragdula*. Also the scutellum of the male is described as carrying a more or less distinct blackish patch laterally of lateral keels, but similar patches can sometimes be observed in males of *smaragdula*. Male pygofer as in Text-fig. 257, anal tube of male as in Text-figs. 258, 259, style as in Text-fig. 260, aedeagus as in Text-figs. 261–263. Ventral aspect of apical part of female abdomen as in Text-fig. 264. Serration of ovipositor saw coarse (Text-fig. 265). Length of macropters 3.9–5.3 mm, of brachypters 3.3–4 mm.

Distribution. Scarce in Denmark, found in EJ, NWJ, F, LFM, and SZ. – In Sweden so far only found in Bl., Karlskrona, Våmö and Gullberna in June, 1972 and 1976 (N. Gyllensvärd). – Not in Norway, nor in East Fennoscandia. – France, England, Poland.

Biology. On *Phragmites*, adults in June–July.

36. *Chloriona glaucescens* Fieber, 1866
Text-figs. 266–273.

Chloriona glaucescens Fieber, 1866b: 522.
Chloriona unicolor J. Sahlberg, 1871: 406 (nec Herrich-Schäffer, 1835).

Very like *Chloriona smaragdula* but the genital segment of the male is light in colour, and both sexes differ by details in their genitalia. Male pygofer as in Text-fig. 266, anal tube of male as in Text-fig. 267, style as in Text-fig. 268, aedeagus as in Text-figs. 269–271. Ventral aspect of apical part of female abdomen as in Text-fig. 272, serration of saw of ovipositor coarse (Text-fig. 273). Length of macropters with wings 4.7–5.7 mm, of brachypters 4.4–5.1 mm.

Distribution. Fairly common in Denmark, found in SJ, EJ, F, LFM, and NEZ. – Comparatively common in southern Sweden, found in Sk., Bl., Hall., Öl., Gtl., and Ög. – In Norway established only in VAy: Mandal, and Rona Mandal, July, 1935 (Soot-Ryen). – In East Fennoscandia this is the commonest *Chloriona* species, found in Al, Ab, N, Oa, Kb, and Om. – Widespread in Europe, also in Kazakhstan and Uzbekistan.

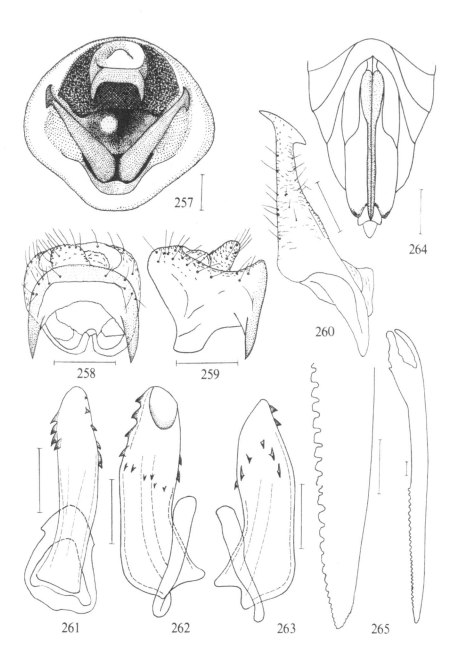

257

258 259 260 264

261 262 263 265

Biology. On *Phragmites*, adults in June–August. Hibernation takes place in the larval stage (Müller, 1957).

37. *Chloriona vasconica* Ribaut, 1934
Plate-fig. 30, text-figs. 274–282.

Chloriona vasconica Ribaut, 1934: 286.

Very like *glaucescens*, differing by male genitalia and by the shape of the lateral lobes of females. One female in my collection (from Ög., Kimstad) belongs to f. *intermedia*, fore wings being twice as long as broad, hind wings vestigial. In this specimen the pygofer is apically fuscous (as described for *dorsata*). Male pygofer as in Text-fig. 275, anal tube of male as in Text-fig. 276, style as in Text-fig. 277, aedeagus as in Text-figs. 278–280. Venter of posterior part of female abdomen as in Text-fig. 281, saw of ovipositor with coarse serration (Text-fig. 282). Length of macropters 4–5 mm, of brachypters 3.5–5 mm.

Distribution. Rare in Denmark, only taken in NWJ: Lund Fjord 8.VII.1974 by E. Bøggild. – Probably not rare in Sweden, so far found in Bl., Karlskrona, Gullberna (N. Gyllensvärd), Ög., Kullerstad and Kimstad (Ossiannilsson), Upl., Djursholm, lake Ösbysjön, and Solna, lake Råstasjön (Ossiannilsson). – Not established in Norway, nor in East Fennoscandia. – Czechoslovakia, France, German D.R. and F.R., England, Hungary, Italy, Poland, s. Russia.

Biology. On *Phragmites*, adults in June and July.

SUBFAMILY CRIOMORPHINAE

Frons in last larval instar and adults as in Stenocraninae. Wing-dimorphism common. Body more or less short and broad. Teeth of post-tibial calcar not broadened. Saw-case of female not shield-shaped.

Genus *Megamelus* Fieber, 1866

Megamelus Fieber, 1866b: 519.
Type-species: *Delphax notula* Germar, 1830, by monotypy.

Text-figs. 257–265. *Chloriona dorsata* Edwards. – 257: male pygofer from behind; 258: male anal tube from behind; 259: male anal tube from the left; 260: genital style; 261: aedeagus, ventral aspect; 262: aedeagus from the right; 263: aedeagus from the left; 264: caudal part of female abdomen from below; 265: saw of ovipositor and apex in higher magnification. Scale: 0.5 mm for 264, 0.1 mm for the rest.

103

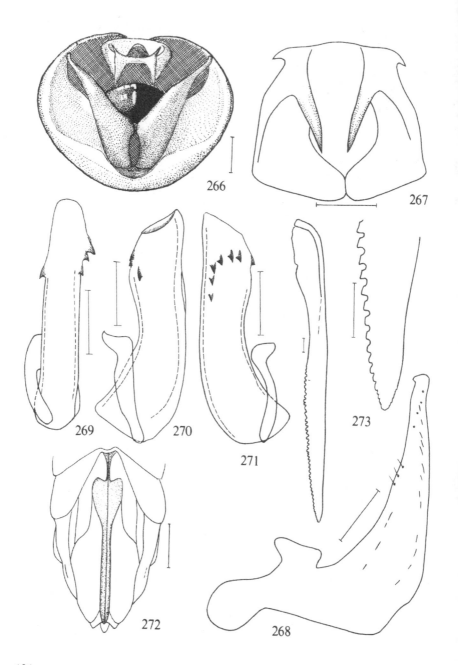

266   267

269   270   271   273

272   268

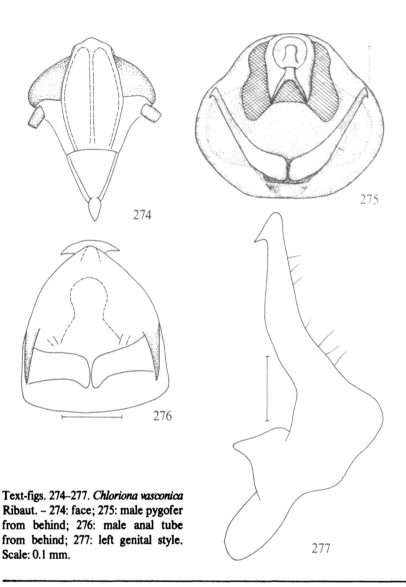

Text-figs. 274–277. *Chloriona vasconica* Ribaut. – 274: face; 275: male pygofer from behind; 276: male anal tube from behind; 277: left genital style. Scale: 0.1 mm.

Text-figs. 266–273. *Chloriona glaucescens* Fieber. – 266: male pygofer from behind; 267: male anal tube from behind; 268: left genital style; 269: aedeagus, ventral aspect; 270: aedeagus from the left; 271: aedeagus from the right; 272: caudal part of female abdomen from below; 273: saw of ovipositor and apex in higher magnification. Scale: 0.5 mm for 272, 0.1 mm for the rest.

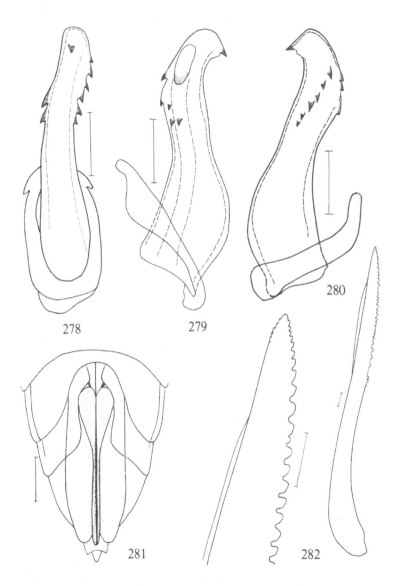

Text-figs. 278–282. *Chloriona vasconica* Ribaut. – 278: aedeagus, ventral aspect; 279: aedeagus from the left; 280: aedeagus from the right; 281: caudal part of female abdomen from below; 282: saw of ovipositor and apex in higher magnification. Scale: 0.5 mm for 281, 0.1 mm for the rest.

First and second antennal segments cylindrical. Frons fairly long and narrow, its median carina sharp right up to its upper end. Median carina of clypeus sharp. Vertex extending considerably in front of eyes (Text-fig. 283). Pronotum with 3 carinae, lateral carinae little divergent, reaching hind border (Text-fig. 283). Distance between median and lateral keel at posterior border of pronotum considerably shorter than length of median carina. First segment of hind tarsi at least at long as 2nd and 3rd segment together. Post-tibial calcar with 13-24 teeth, apical tooth absent. In North Europe one wing-dimorphous species.

38. *Megamelus notula* (Germar, 1830)
Plate-fig. 17, text-figs. 283-291.

Delphax notula Germar, 1830: 57.
Delphax truncatipennis Boheman, 1847b: 266.
Megamelus brevifrons Reuter, 1880: 235.

Fore wings of the brachypterous form (Plate-fig. 17) apically truncate, covering the basal abdominal terga only. Yellowish white to yellowish brown. Frons long and narrow with almost straight side-margins, broadest just above clypeus, fuscous with some small lighter spots and a light transverse band above clypeus, or mottled in brownish to yellowish white. Carinae of head light. Vertex as well as pro- and mesonotum laterally of side carinae mainly brownish. Tegulae light. Fore wings of brachypters usually with a dark longitudinal band laterally of the claval suture. This band may be reduced or, on the contrary, extending over the major part of the wing surface. There is also a dark spot in the lateral apical angle of the fore wing. In some specimens, especially females, the fore wings are entirely light without markings, or largely dark. The markings of the fore body are continued by two broad dark longitudinal bands on the abdominal terga. In some individuals the dark pigment extends over almost the entire dorsum. In the macropterous form the mesonotum is usually wholly dark, the fore wings are transparent with partly dark veins and an oblong dark patch in apex of clavus. Underneath side and legs more or less dark. Pygofer of male (Text-figs. 284, 285) conspicuously large, on each side with a shell-shaped projection, caudally with a pair of erect processes arising from ventral border, genital phragm also with a pair of thin erect processes visible between appendages of anal tube. Anal tube of male (Text-figs. 286, 287) with two thin sharp-pointed appendages. Styles as in Text-fig. 288, aedeagus (Text-fig. 289) very long and thin, with a thin appendage arising from near basis. Lateral lobes of female very broad, basally usually overlapping one another (Text-fig. 290), genital scale distinct, broad (Text-fig. 291). Overall length of macropters 3.9-5 mm, of brachypters 2.25-4.2 mm.

Distribution. Common and widespread in Denmark and Sweden (Sk.-Nb.), as well as in East Fennoscandia (Al, Ab – ObN, Ks, also Vib and Kr). – In Norway found from AK to Ry. – Widespread in Europe, established also in Azerbaijan, Kazakhstan, w. Siberia, Mongolia, and Japan.

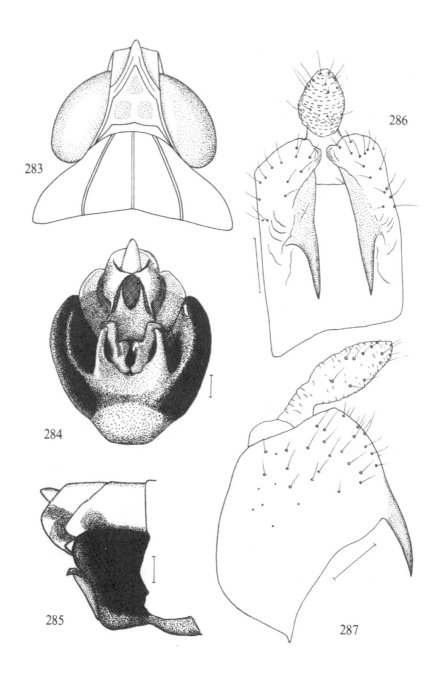

283

286

284

285

287

Biology. Often very abundant on *Carex* in wet biotopes. Host-plant *Carex riparia* (Müller, 1951). Adults from July on, hibernation in the adult stage.

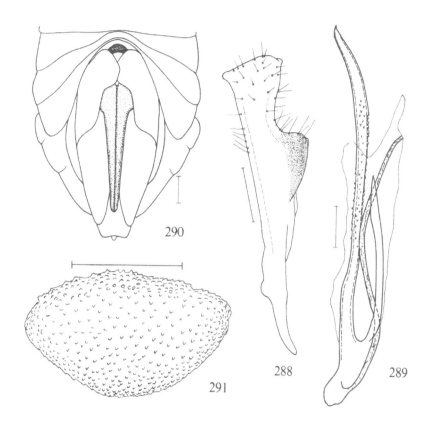

Text-figs. 288–291. *Megamelus notula* (Germar). – 288: left genital style from behind; 289: aedeagus; 290: caudal part of female abdomen from below; 291: genital scale from below. Scale: 0.25 mm for 290, 0.1 mm for the rest.

Text-figs. 283–287. *Megamelus notula* (Germar). – 283: head and prothorax from above; 284: male pygofer from behind; 285: male pygofer from the right; 286: male anal tube from behind; 287: male anal tube from the left. Scale: 0.1 mm.

Genus *Unkanodes* Fennah, 1956

Unkanodes Fennah, 1956: 474.
Type-species: *Unkana sapporona* Matsumura, 1935, by original designation.
Elymodelphax W. Wagner, 1963: 167.
Type-species: *Liburnia excisa* Melichar, 1898, by original designation.

Body comparatively slender. Head little narrower than pronotum. Vertex longer than broad, its basal width not exceeding width of an eye, shallowly rounded at apical margin. Carinae of vertex and frons distinct. Frons much longer than broad, its median carina forked at junction with vertex. Antennae cylindrical, first segment 2-2.5 times as long as broad, at least half as long as second. Lateral carinae of pronotum almost straight, not reaching hind margin and not in line with mesonotal carinae. Teeth of post-tibial calcar well developed. Pygofer of male with a lateral incision. In Denmark and Fennoscandia one species.

39. *Unkanodes excisa* (Melichar, 1898)
Text-figs. 292–304.

Liburnia excisa Melichar, 1898: 67.
Liburnia elymi Jensen-Haarup, 1915: 139, 144.

Sand-coloured. Frons narrow, almost parallel-sided, of a uniform pale yellowish colour, sometimes darker marbled between carinae. Genae and clypeus uniformly pale yellowish or with more or less extended fuscous markings. Dorsum of fore body with a whitish median longitudinal band. Abdomen of male blackish brown with light spots and light segment borders, hind margin of pygofer yellowish white. Abdomen of female yellowish white with diffuse black and brownish spots. Fore wings of brachypters hyaline, well twice as long as broad, reaching near apex of abdomen. Fore wings of macropters by 2/5 longer than abdomen, veins fuscous towards apex. Male pygofer as in Text-figs. 292, 293. Genital phragm medially on ventral border with a pair of upwards directed hooks (Text-figs. 294, 295). Anal tube of male (Text-figs. 296, 297) with a pair of strong curved pointed appendages. Styles as in Text-figs. 298, 299, aedeagus as in Text-figs. 300–302. Venter of posterior part of female abdomen as in Text-fig. 303. Genital scale (Text-fig. 304) large, broad. Overall length of macropters 4–4.7 mm, of brachypters 2.2–3.5 mm.

Distribution. Scarce in Denmark, found in SJ, EJ, F, and NEZ, and in Sweden (Sk.,

Text-figs. 292–299. *Unkanodes excisa* (Melichar). – 292: male pygofer from behind; 293: male pygofer from the right; 294: genital phragm from behind; 295: genital phragm from the left; 296: male anal tube from behind; 297: male anal tube from the left; 298: left genital style from outside; 299: left genital style from behind. Scale: 0.1 mm.

110

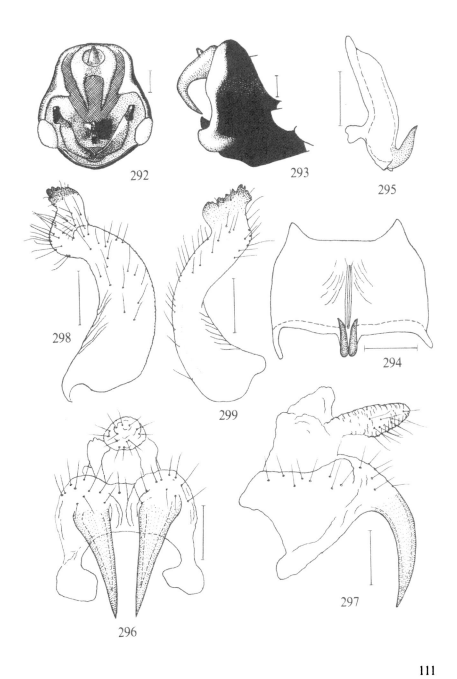

292   293

295

298

299

294

296   297

Hall., Gtl., G. Sand., Hls., Vb., Nb.). – So far not established in Norway. – Comparatively rare in East Fennoscandia, found in Al, N, Ka, St, Om, ObN, and Vib. – German D.R. and F.R., Poland, Estonia, N. Russia, Ukraine, Kurile Isles.

Biology. On *Elymus arenarius* at seashores, adults in May–August.

Genus *Megadelphax* W. Wagner, 1963

Megadelphax W. Wagner, 1963: 167.
Type-species: *Delphax sordidula* Stål, 1853, by original designation.

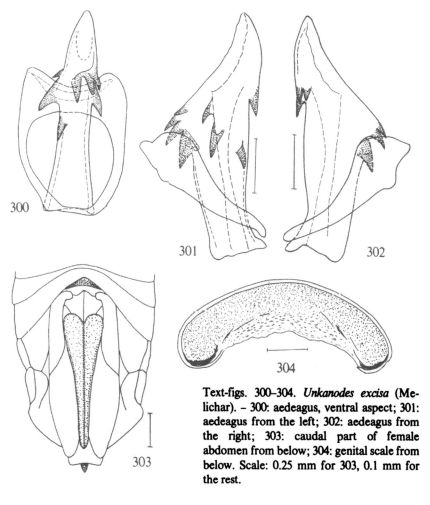

Text-figs. 300–304. *Unkanodes excisa* (Melichar). – 300: aedeagus, ventral aspect; 301: aedeagus from the left; 302: aedeagus from the right; 303: caudal part of female abdomen from below; 304: genital scale from below. Scale: 0.25 mm for 303, 0.1 mm for the rest.

Vertex as long as broad or distinctly longer, carinae distinct. Frons about twice as long as broad, broadest between lower margin of eyes, sides beneath eyes more or less straight. Carinae of pronotum as in *Unkanodes*. Teeth of post-tibial calcar small, 14–21 in number. In Fennoscandia two species.

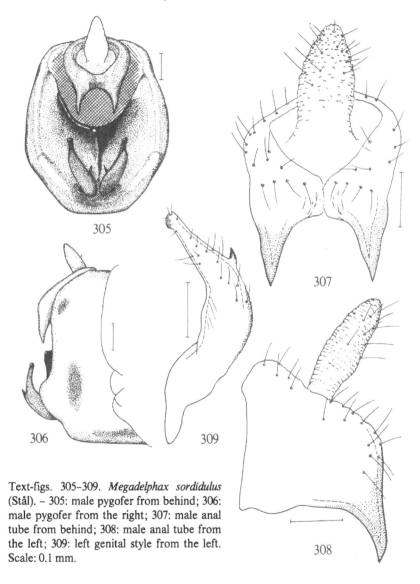

Text-figs. 305–309. *Megadelphax sordidulus* (Stål). – 305: male pygofer from behind; 306: male pygofer from the right; 307: male anal tube from behind; 308: male anal tube from the left; 309: left genital style from the left. Scale: 0.1 mm.

113

Key to species of *Megadelphax*

1 Large species, brachypters 3.2–4.4 mm. Fore wings of brachypters twice as long as broad, apically rounded. Styles of male on inside each with a small sharp upturned tooth (Text-fig. 309). Lateral lobes of female very broad (Text-fig. 313)

40. *sordidulus* (Stål)

– Smaller species, brachypters 2.25–3 mm. Fore wings of brachypters 1.4–1.8 times as long as broad, apically truncate. Styles of male without a tooth. Lateral lobes of female not especially broad (Text-fig. 324)   41. *haglundi* (J. Sahlberg)

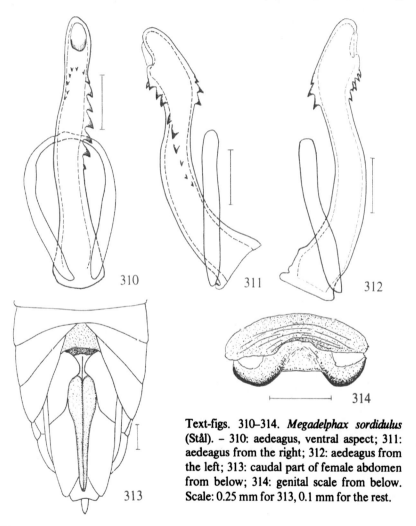

Text-figs. 310–314. *Megadelphax sordidulus* (Stål). – 310: aedeagus, ventral aspect; 311: aedeagus from the right; 312: aedeagus from the left; 313: caudal part of female abdomen from below; 314: genital scale from below. Scale: 0.25 mm for 313, 0.1 mm for the rest.

114

40. Megadelphax sordidulus (Stål, 1853)
Plate-figs. 32, 33, text-figs. 305-314.

Delphax sordidula Stål, 1853: 174.

Brownish yellow or sordid yellow. Carinae of frons indistinctly or not at all bordered with fuscous. Dorsum of fore part of body with a lighter central longitudinal stripe. Abdomen of male black with longitudinal rows of light spots. Abdomen of female sordid yellow with more or less indistinctly limited darker spots along sides, or even largely dark. Fore wings of brachypters pale, transparent, about twice as long as broad, apically rounded, covering about 2/3 of abdomen. Fore wings of macropters 1 1/2-1 2/3 times as long as abdomen, hyaline, veins yellowish, fuscous towards apex. Pygofer of male as in Text-figs. 305, 306, anal tube of male as in Text-figs. 307, 308, style (Text-fig. 309) apically truncate, on inside with a small sharp tooth. Aedeagus as in Text-figs. 310-312. Lateral lobes of female very broad (Text-fig. 313), genital scale as in Text-fig. 314. Overall length of macropters 4.8-5.1 mm, of brachypters 3.2-4.4 mm.

Distribution. So far not found in Denmark, nor in Norway. – Comparatively common in Central Sweden, found in Gtl., Ög., Sdm., Upl., Dlr., Med., Jmt., Vb. – Common in East Fennoscandia, Al, Ab N – Ok, also in Kr. – Not in Great Britain, nor in the Pyrenean Peninsula, otherwise widespread in Europe, also found in Algeria, Tunisia, Kazakhstan, m. Siberia, and Mongolia.

Biology. Locally abundant on grass meadows, leys and cereal fields, adults in June–August. Univoltine. Hibernation takes place in the larval stage (Müller, 1957), in Finland usually in instars II and III (Raatikainen, 1970).

Economic importance. Megadelphax sordidulus is a vector of Phleum green stripe virus (PGSV). It is occasionally important in leys (Raatikainen, l. c., Heikinheimo & Raatikainen, 1976).

41. Megadelphax haglundi (J. Sahlberg, 1871) comb. n.
Text-figs. 315-326.

Liburnia Haglundi J. Sahlberg, 1871: 427.

Carinae of head ivory-white. Vertex anteriorly black, caudal impressions orange-coloured. Carinae of frons more or less broadly margined with black or interspaces almost entirely black. Pro- and mesonotum orange or yellow, carinae and a broad median band on scutellum white. Thoracal venter of male partly black. Abdomen of male black with a median and some lateral longitudinal series of brownish yellow spots, abdominal segments VII and VIII dorsally broadly light. Pygofer of male black, dorsally light, anal tube and anal style light. Thoracal venter and abdomen of brachypterous female entirely yellow, those of macropterous female yellow with diffuse dark markings. Fore wings of macropterous female about twice as long as abdomen (one specimen seen). I have not seen a macropterous male. Teeth of post-tibial calcar (Text-

115

fig. 315) small but distinct. Male pygofer as in Text-figs. 316 and 317. Genital phragm of male above lower opening with a spinose elevated median carina (Text-fig. 318). Anal tube of male (Text-figs. 319, 320) with two fairly long, parallel, pointed appendages. Styles (Text-fig. 321) comparatively short, apex blunt. Aedeagus as in Text-figs.

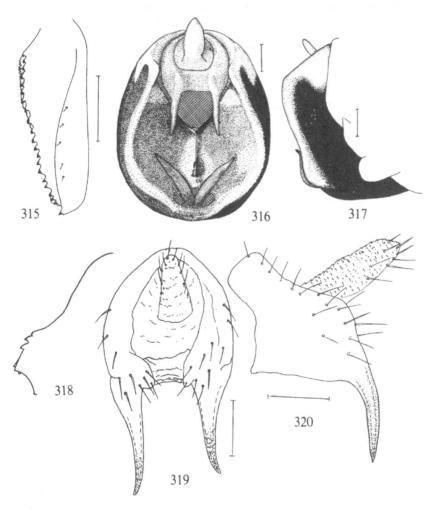

Text-figs. 315–320. *Megadelphax haglundi* (J. Sahlberg). – 315: post-tibial calcar; 316: male pygofer from behind; 317: male pygofer from the right; 318: lateral outline of median carina of genital phragm, from the right; 319: male anal tube from behind; 320: male anal tube from the left. Scale: 0.1 mm.

116

322–324. Venter of posterior part of female abdomen as in Text-fig. 325, genital scale (Text-fig. 326) fairly large. Length of brachypters 2.25–3 mm, macropterous female (with wings) 3.8 mm.

Distribution. Sweden: Ög., Kimstad (Haglund), Norrköping, Alsäter (A. Tullgren), Skärkind, Karlslund (Tullgren), Askeby 1932–34 (Ossiannilsson), Stjärnorp (Ossiannilsson); Upl., Vallentuna, vicinity of Fagermoda 1969 (Ossiannilsson). – Outside Sweden only found in Moravia and Bohemia (Dlabola, 1955).

Biology. The ecology of *M. haglundi* has not been subjected to a study so far. The only locality where this very rare species has been found in some numbers is the one in

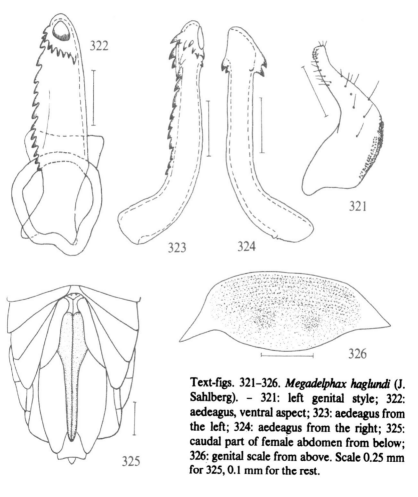

Text-figs. 321–326. *Megadelphax haglundi* (J. Sahlberg). – 321: left genital style; 322: aedeagus, ventral aspect; 323: aedeagus from the left; 324: aedeagus from the right; 325: caudal part of female abdomen from below; 326: genital scale from above. Scale 0.25 mm for 325, 0.1 mm for the rest.

117

Askeby. In this place e. g. *Jassargus distinguendus, Turrutus socialis,* and *Muellerianella fairmairei* were also collected. Our specimens were captured in June and July (9.VI.–26.VII.).

Genus *Laodelphax* Fennah, 1963

Laodelphax Fennah, 1963: 15.
Type-species: *Delphax striatella* Fallén, 1826, by original designation.
Callidelphax W. Wagner, 1963: 167.
Type-species: *Delphax striatella* Fallén, 1826, by original designation.

"Delicately built. Vertex quadrate, as long as broad, slightly narrower than eye, anteriorly truncate, carinae distinct; frons about twice as long as broad; rostrum just surpassing mesotrochanters; lateral pronotal carina concave, incomplete, legs long and slender; calcar tectiform, many toothed. Pygofer very short dorsally, longer and convex ventrally, lateral margins not entire, no medioventral process or notch; diaphragm broad, dorsally shallowly excavate". (Fennah, 1. c.). Post-tibial calcar with 10–15 teeth, apical tooth punctiform or missing. Only one species.

42. *Laodelphax striatellus* (Fallén, 1826)
Text-figs. 327–334.

Delphax striata Fallén, 1806: 129 (nec Fabricius, 1794).
Delphax striatella Fallén, 1826: 75.
Liburnia marginata Haupt, 1935: 42 (nec Fabricius, 1794).
Calligypona marginata Ossiannilsson, 1946c: 55 (nec Fabricius, 1794).

Wing-dimorphous, usually macropterous. Male black. Carinae of frons and clypeus, pronotum (except a large spot behind each eye) and apex of scutellum, normally whitish. Vertex and tegulae yellowish. Fore wings with a black streak in apex of clavus. Antennae and legs yellowish. Fore wings of brachypters hyaline – yellowish, about 2.4 x as long as broad, little longer than abdomen, apically rounded, veins concolorous or fuscous. Fore wings of macropters nearly twice as long as abdomen, hyaline with veins darkened towards apex. Terga of 1st and 2nd abdominal segments orange or yellowish. Scutellum of female with a broad, light, median longitudinal band, venter of abdomen partly light. Hind margin of male pygofer (Text-figs. 327, 328) with a lateral incision, anal tube of male with a pair of short pointed appendages situated widely apart (Text-figs. 329, 330). Styles short, apically truncate (Text-fig. 331). Aedeagus (Text-figs. 332, 333) slightly curved, without appendages. Venter of posterior part of female abdomen as in Text-fig. 334, lateral lobes basally with an angular projection. Length of brachypters 1.75–3 mm, of macropters (with wings) 3–4.75 mm.

Distribution. So far not established in Denmark. – Only once found in Norway: HEs Løten 29.VIII.1961, one male (H. Holgersen). – Comparatively scarce in Sweden, found in Sk., Bl., Sm., Öl., Gtl., Ög., Vg., Sdm., Upl., Dlr., Hls. – Rare in East

118

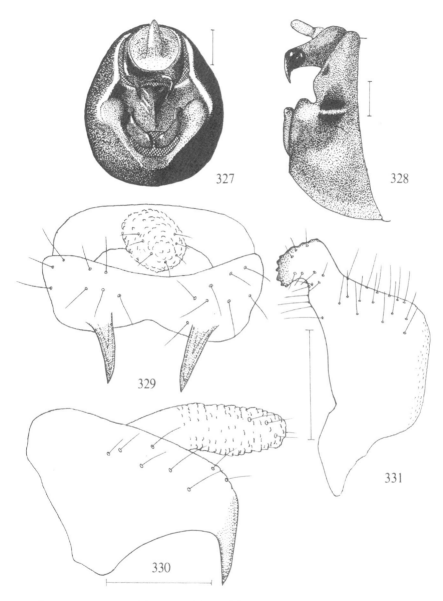

Text-figs. 327–331. *Laodelphax striatellus* (Fallén). – 327: male pygofer from behind; 328: male pygofer from the right; 329: male anal tube from behind; 330: male anal tube from the left; 331: left genital style. Scale: 0.1 mm.

Fennoscandia, established in Al, Ab, ObN, LkE, Vib, and Kr. – Widespread in the Palaearctic region, found also in the Oriental region.

Biology. Rarely abundant in our territory. On grasses in cultivated fields, also in both wet and dry grass meadows. Adults in May–September. Two generations in Sweden (M. Azrang in litt.). In Siberia there are two, in Japan 4–7 generations. Hibernation takes place in larval stages.

Economic importance. *Laodelphax* is an important virus vector, in Sweden transmitting cereal tillering disease on barley and oats (Lindsten & Gerhardsen, 1971). Other cereal diseases are transmitted by this species in Russia, Siberia, and Japan. Its importance as a virus vector in Sweden is limited by its comparative rarity and low abundance in this country.

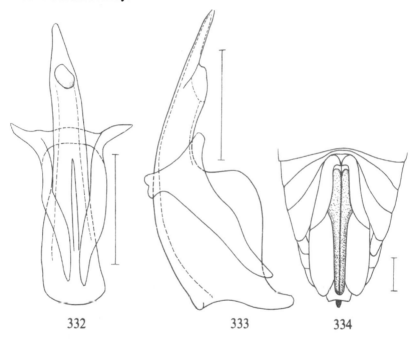

332   333   334

Text-figs. 332–334. *Laodelphax striatellus* (Fallén). – 332: aedeagus, ventral aspect; 333: aedeagus from the left; 334: caudal part of female abdomen from below. Scale: 0.25 mm for 334, 0.1 mm for the rest.

Genus *Paraliburnia* Jensen-Haarup, 1917

Paraliburnia Jensen-Haarup, 1917: 1.
Type-species: *Paraliburnia jacobseni* Jensen-Haarup, 1917, by original designation.

120

Vertex as long as broad or indistinctly longer. First and second antennal segments short. Frons almost parallel-sided, about twice as long as broad. Calcar with 24–30 marginal teeth in a straight row. Posterior margin of male pygofer in lateral aspect without incision. Appendages of anal tube in male parallel, widely apart. Genital styles of almost same width throughout, or a little broader at apex, curved, apically somewhat convergent. Two species.

Key to species of *Paraliburnia*

1 Fore wing of brachypters more than twice as long as broad. Legs longer: index hind tibia + hind tarsus: width of head (with eyes) = 2.58–2.65 (average 2.64). Genital styles of male apically distinctly widened, but without a distinct outwards directed projection (Text-fig. 339)   43. *adela* (Flor)
- Fore wing of brachypters about 1 1/2 times as long as broad. Legs shorter: index hind tibia + hind tarsus: width of head = 2.25–2.52 (average 2.40). Styles apically widened, with a distinct outwards directed projection (Text-fig. 347) 44. *clypealis* (J. Sahlberg)

43. *Paraliburnia adela* (Flor, 1861)
Text-figs. 335–342.

Delphax adela Flor, 1861: 63.
Delphax concolor Fieber, 1866: 529.
Paraliburnia jacobseni Jensen-Haarup, 1917: 2.

Median carina of frons sharp but nearly evanescent on junction to vertex. Wing trimorphous. Head, pronotum, scutellum and legs of brachypterous male dirty brownish yellow, abdomen black. Frons between carinae may be fuscous, then the carinae are lighter; in light specimens frons and carinae are concolorous. Fore wings of brachypters sordid yellow, transparent, nearly as long as abdomen, apically rounded. Fore wings of macropters about twice as long as abdomen. Scutellum of macropters dark brown or blackish. Female usually brachypterous, light yellow-brown, abdomen often darker laterally, fore wings brownish or smoky tinged, reaching more than half length of abdomen. Male pygofer as in Text-figs. 335 and 336, anal segment of male as in Text-figs. 337, 338, genital style as in Text-fig. 339, aedeagus as in Text-figs. 340, 341. Venter of posterior part of female abdomen as in Text-fig. 342. Length of brachypters 2.7–4 mm, of macropters 4.25–4.8 mm, of intermediate form (according to Le Quesne, 1960) 3.7 mm.
    Note. I have not seen females of this species and statements concerning them have been compiled from the literature. Text-fig. 342 was copied after Fig. 95H in Vilbaste (1971).

    Distribution. Scarce in Denmark, established in WJ, LFM, SZ, NEZ, and B. – Very rare in Sweden and East Fennoscandia. Bo Tjeder found one male in Boh., Ljung,

121

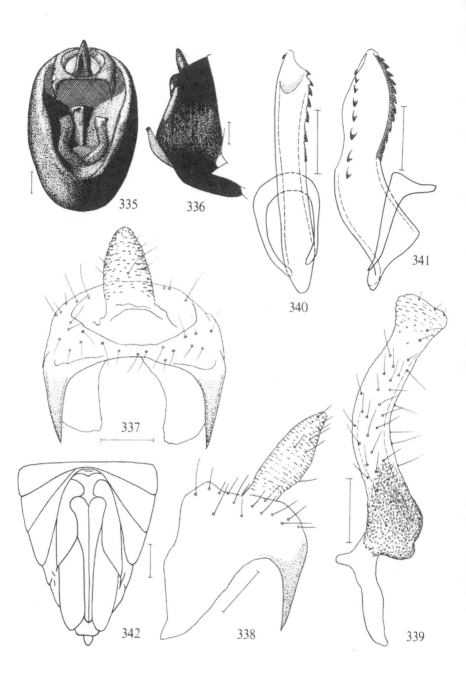

335 336

340

341

337

342 338 339

122

Anfasteröd 26.VI.1946. K. Lindsten captured another male in Ög., Alvastra, vicinity of Dags mosse, in February, 1973. Ossiannilsson collected in all four males in Upl., Uppsala, Ultuna and Kungsängen 26-28.VI.1952. Håkan Lindberg found *adela* in Ab, Lojo, 24.VI.1942. – It has not been established in Norway so far. – Absent from the Mediterranean area, otherwise widespread in Europe, also found in Kazakhstan and Siberia.

Biology. On wet meadows, on *Phalaris arundinacea* (Strübing, 1956). Also *Glyceria* spp. have been recorded as host-plants (Le Quesne, 1960, Linnavuori, 1969, Vilbaste, 1971). The find by Lindsten in February indicates that hibernation in our climatic conditions can take place in the adult stage.

44. Paraliburnia clypealis (J. Sahlberg, 1871)
Text-figs. 343–353.

Liburnia clypealis J. Sahlberg, 1871: 454.
Calligypona litoralis Ossiannilsson, 1944: 16 (nec Reuter, 1880).

Wing dimorphous, usually brachypterous. Frons widest at level of lower margin of compound eye, median carina as in *adela*. Anterior part of dorsum and fore wings in brachypterous male dirty yellow, frons and clypeus fuscous or blackish, carinae brownish yellow. Apex of first antennal segment and basis of second segment dark. Venter of thorax black-spotted, abdomen black, legs dirty yellow or yellowish brown. Brachypterous female yellowish brown to chestnut brown, frons more or less distinctly mottled with fuscous. Fore wings of brachypters (both sexes) about 1 1/2 times as long as broad, apically rounded, covering half abdomen, veins apically darker. Macropterous female dirty brownish, prothorax and carinae of head lighter, scutellum with a broad median band and lateral angles dirty yellow, legs also dirty yellow. Wings of macropters about twice as long as abdomen. Male pygofer as in Text-figs. 343, 344, anal tube of male with two short pointed appendages (Text-figs. 345, 346), styles as in Text-fig. 347, aedeagus as in Text-figs. 348–350. Ventral aspect of posterior part of female abdomen as in Text-fig. 351. Genital scale (Text-figs. 352, 353) small, caudally attached to two irregularly lump-shaped bodies. Overall length of macropters 3.7–4.1 mm, of brachypters 2.0–3.4 mm.

Distribution. So far not established in Denmark, nor in Norway. – In Sweden only found in Upl., Djursholm, lake Ösbysjön, and Solna, lake Råstasjön, Upsala-Näs, Ytternäs, and Vallentuna, Grindstugan (Ossiannilsson). – In East Fennoscandia found in Ab,

Text-figs. 335–342. *Paraliburnia adela* (Flor). – 335: male pygofer from behind; 336: male pygofer from the right; 337: male anal tube from behind; 338: male anal tube from the left; 339: left genital style; 340: aedeagus, ventral aspect; 341: aedeagus from the right; 342: caudal part of female abdomen from below. Scale: 0.25 mm for 342, 0.1 mm for the rest. (342 after Vilbaste, 1971).

123

Raisio (Linnavuori); Kb, Hammaslahti (Kontkanen), also in Vib, Raivola and Keksholm (J. Sahlberg). – England, German D.R. and F.R., Czechoslovakia, n. Russia.

Biology. In tufts of *Calamagrostis canescens* in wet meadows etc., locally fairly abundant. Adults in June–July.

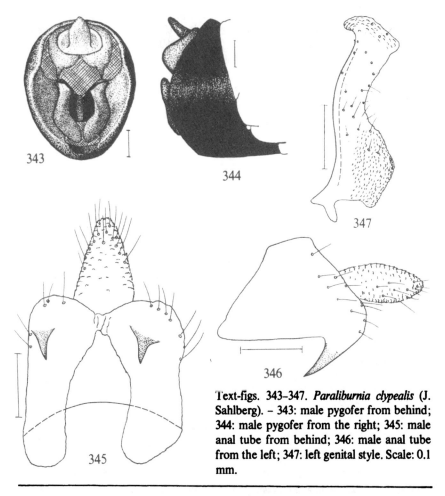

Text-figs. 343–347. *Paraliburnia clypealis* (J. Sahlberg). – 343: male pygofer from behind; 344: male pygofer from the right; 345: male anal tube from behind; 346: male anal tube from the left; 347: left genital style. Scale: 0.1 mm.

Text-figs. 348–353. *Paraliburnia clypealis* (J. Sahlberg). – 348: aedeagus, ventral aspect; 349: aedeagus from the right; 350: aedeagus from the left; 351: caudal part of female abdomen from below; 352: genital scale; 353: genital scale with attached lump-shaped bodies. Scale: 0.25 mm for 351, 0.1 mm for the rest.

124

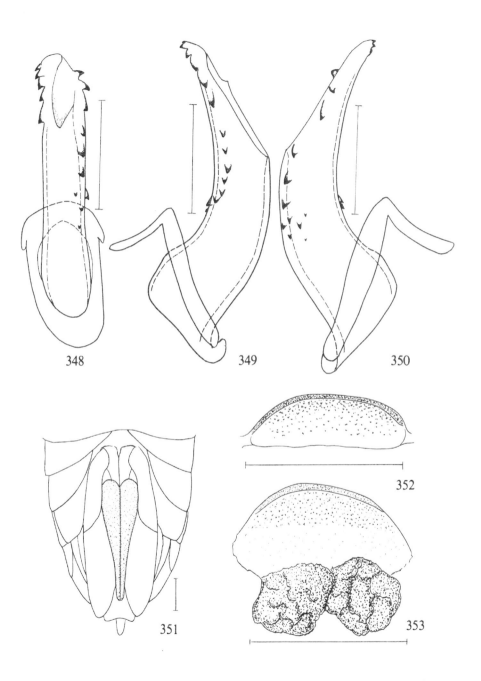

348 349 350 351 352 353

Genus *Hyledelphax* Vilbaste, 1968

Hyledelphax Vilbaste, 1968: 71.

Type-species: *Delphax elegantula* Boheman, 1847, by original designation.

Vertex distinctly but not much longer than broad. Frons about twice as long as broad, sides evenly curved. Carinae of vertex and frons distinct, median carina of frons forked on junction to vertex. Lateral carinae of pronotum not reaching hind margin, apically curved outwards. Lateral carinae of scutellum distinct, diverging caudally. Fore wings of brachypters little longer than broad, apically truncate. Post-tibial calcar comparatively small, with about 14 marginal teeth. Male pygofer laterally with a broad and

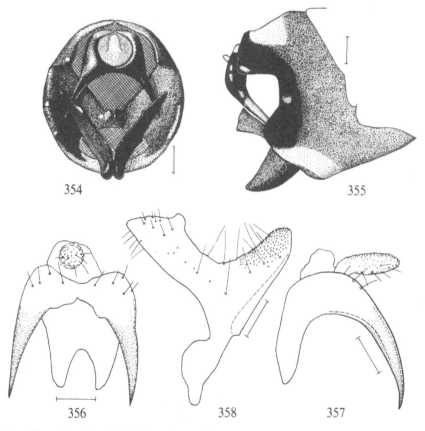

Text-figs. 354–358. *Hyledelphax elegantulus* (Boheman). – 354: male pygofer from behind; 355: male pygofer from the right; 356: male anal tube from behind; 357: male anal tube from the left; 358: left genital style. Scale: 0.1 mm.

126

deep incision (Text-fig. 355). Appendages of male anal tube long, widely apart. Styles broad (Text-fig. 358), with a basal process directed backwards. Genital scale of female (Text-fig. 364) comparatively large, bilobate, black. Carinae of frons in nymphs more or less straightly converging, almost uniting beneath. One species.

45. *Hyledelphax elegantulus* (Boheman, 1847)
Plate-fig. 16, text-figs. 354–364.

Delphax elegantula Boheman, 1847a: 63.

Frons and anterior part of vertex in brachypterous male black between greyish white

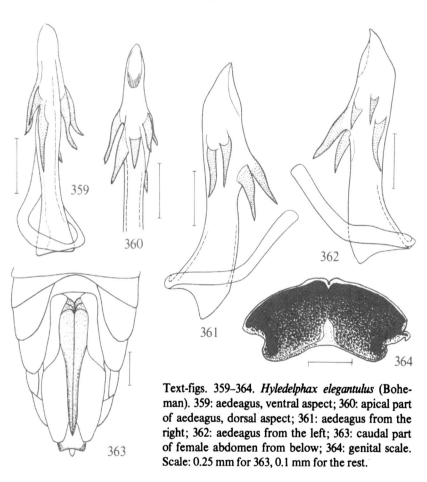

Text-figs. 359–364. *Hyledelphax elegantulus* (Boheman). 359: aedeagus, ventral aspect; 360: apical part of aedeagus, dorsal aspect; 361: aedeagus from the right; 362: aedeagus from the left; 363: caudal part of female abdomen from below; 364: genital scale. Scale: 0.25 mm for 363, 0.1 mm for the rest.

127

carinae. Clypeus and caudal part of vertex yellowish brown, alternatively the latter may be white. Pronotum white except for a broad black spot behind each eye and sometimes also two diffuse dark spots more medially at fore border. Scutellum black, apex whitish. Fore wings of brachypterous males transparent, greyish white, with a yellow-fuscous tinge here and there on veins. Ventral side of thorax and abdomen largely black, abdomen dorsally with a usually broad yellowish brown median longitudinal band, often on posterior segments only. In the macropterous male the dark markings of pronotum are more extended than in the brachypterous one. Fore wings of macropters transparent, greyish to whitish with fuscous veins. The female is yellowish brown or sordid yellow with carinae of frons whitish or pale yellowish bordered with black; abdomen usually with longitudinal rows of fuscous spots along sides. Wings more or less as in male. Male pygofer as in Text-figs. 354, 355, anal tube of male as in Text-figs. 356, 357, style as in Text-fig. 358, aedeagus as in Text-figs. 359–362. Venter of caudal part of female abdomen as in Text-fig. 363, genital scale as in Text-fig. 364. Overall length of macropters 3.2–4 mm, of brachypters 2.0–3.4 mm.

Distribution. Common in Denmark, found in most provinces. – Common also in Sweden, Sk. – Lu. Lpm. – In Norway up to TRy and TRi. – Common in southern and Central East Fennoscandia, scarcer in the north, recorded up to Le. – Widespread in Europe, also reported from Tunisia, Kazakhstan, and Mongolia.

Biology. A heliophilous species, apparently associated with *Deschampsia flexuosa* on moors and heaths with *Calluna*, also in wood glades with *Vaccinium* etc. Hibernation in larval stage (Strübing in Müller, 1957, Remane, 1958). Adults in May–August.

Genus *Megamelodes* Le Quesne, 1960

Megamelodes Le Quesne, 1960: 1.
Type-species: *Delphax quadrimaculatus* Signoret, 1865, by original designation.

Lateral carinae of pronotum divergent, reaching hind margin of pronotum, distance between their posterior ends much greater than length of median carina. Vertex less than one and a quarter times as long as broad. Antennae short, basal segment just longer than half length of second. Post-tibial calcar with 14–17 small marginal teeth not arranged in a straight row. Genitalia of male simple without accessory lobes or outgrows of wall of pygofer. Styles long, apices acute, curved towards median line. In North Europe one species.

46. *Megamelodes quadrimaculatus* (Signoret, 1865)
Text-figs. 365–369.

Delphax quadrimaculatus Signoret, 1865: 130.
Liburnia fieberi Scott, 1870: 70.

Vertex as long as broad. Frons twice as long as broad, carinae distinct, side margins moderately convex. Frons dark brown with lighter transverse spots, lower margin pale. Colour of body varying between light brown and almost black, often darker at sides. Fore wings of brachypters apically truncate with rounded angles, about as long as broad, not quite half as long as abdomen in male, 1/3–1/4 as long as abdomen in female, dark brown, on apical margin with two white spots separated by a black spot. Abdomen of male blackish brown, with three more or less distinct longitudinal series of light markings. Abdomen of female yellowish brown, laterally broadly darker, with three more or less distinct lines of pale spots. Scutellum of macropterous females black, fore wings pale with a dark patch in clavus just basally of apex of anal vein. Male pygofer as in Text-fig. 365, style as in Text-fig. 366, aedeagus as in Text-figs. 367–369. Overall length of macropters 3.9–4.1 mm, of brachypters 2.2–3.0 mm.

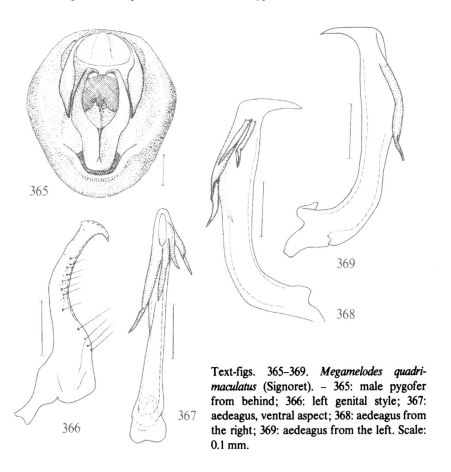

365

366

367

369

368

Text-figs. 365–369. *Megamelodes quadrimaculatus* (Signoret). – 365: male pygofer from behind; 366: left genital style; 367: aedeagus, ventral aspect; 368: aedeagus from the right; 369: aedeagus from the left. Scale: 0.1 mm.

Distribution. Rare in Denmark, only found in EJ: Skovbakken, Randers 13. & 24.VII.1878 by O. Jacobsen. – Not found in Sweden, Norway, and East Fennoscandia. – Austria, Belgium, France, German D.R. and F.R., England, Ireland, Hungary, Italy, Netherlands, and Madeira.

Biology. On very moist meadows, near bases of rushes *(Juncus)*, adults in January–November (Le Quesne, 1960).

Genus *Calligypona* J. Sahlberg, 1871

Calligypona J. Sahlberg, 1871: 408.
Type-species: *Calligypona albicollis* J. Sahlberg, 1871, by monotypy.

Body large, robust. Vertex a little longer than broad. Frons broadest in its lower half, length 2 1/3 times maximal width, median carina distinct, obsolescent on transition to vertex. Antennae cylindrical, second segment almost twice as long as first. Carinae of pronotum curved outwards, not reaching hind border. Legs long: hind tibia + tarsus about 3 times width of head. Fore wings of brachypters 1.7–1.8 times as long as broad, apically obliquely cut off with rounded corners. Post-tibial calcar almost as long as 1st hind tarsal segment, with about 25 marginal teeth arranged in disorder along posterior margin. Styles 5 times as long as broad, with parallel side margins, erect, basally with a backwards directed tooth (Text-fig. 373). Genital scale of female (Text-fig. 377) approximately square with an anterior incision. One species.

47. *Calligypona reyi* (Fieber, 1866)
    Text-figs. 370–377.

Delphax Reyi Fieber, 1866b: 527.
Calligypona albicollis J. Sahlberg, 1871: 409.

Brachypterous male: frons between carinae, anterior part of vertex, antennae and scutellum blackish brown, as well as major part of thoracal venter. Pronotum white, fore wings brownish yellow, abdominal terga I and II orange, remaining part of abdomen blackish brown except for median part of distal segments and lateral spots which are whitish yellow. Pygofer as in Text-figs. 370, 371, anal tube as in Text-fig. 372, style as in Text-fig. 373, aedeagus as in Text-figs. 374, 375. Brachypterous female brownish yellow, pronotum whitish yellow, frons fuscous between lighter carinae, venter with fuscous markings. Ventral aspect of posterior part of female abdomen as in Text-fig. 376, genital scale as in Text-fig. 377. Length of brachypterous male 2.9–3.2

Text-figs. 370–377. *Calligypona reyi* (Fieber). – 370: male pygofer from behind; 371: male pygofer from the right; 372: male anal tube from the left; 373: left genital style; 374: aedeagus, ventral aspect; 375: aedeagus from the right; 376: caudal part of female abdomen from below; 377: genital scale. Scale: 0.25 mm for 376, 0.1 mm for the rest.

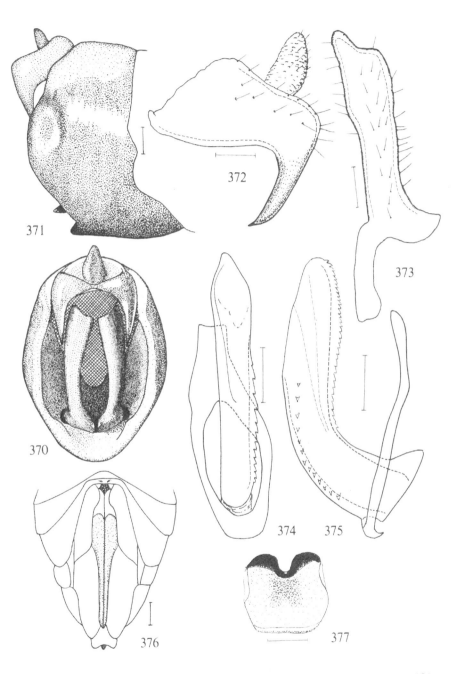

371

372

373

370

374 375

376

377

mm., of brachypterous female 3.6–4.0 mm. Macropterous female (translation from Lindberg, 1932): mesonotum twice as long as pronotum. About one-third of fore wings reaching distally of abdominal apex. Membrane of fore wing one-third shorter than corium. Fore wings transparent, yellowish brown, hind margin distally of claval apex, costal and apical margins blackish.

Distribution. Rare in Denmark. SJ: Høruphav 12.VII.1896 (Wüstnei). – Rare in Sweden, found in Sm. (type of *albicollis*, collector Boheman according to Sahlberg, 1871, Haglund according to Lindberg, 1932); Bl., Åryd (Gyllensvärd); Öl., Hornsjön (A. Jansson); Ög., Rystad, Bjursholmen (Ossiannilsson), Ö. Skrukeby (Ossiannilsson), Ö. Harg (Ossiannilsson). – Not found in Norway. – East Fennoscandia: rare in coastal districts, established in Al, N, and Oa. – Otherwise widespread in Europe, also in Tadzhikistan, Uzbekistan, and Mongolia.

Biology. On *Juncus, Scirpus lacustris* and *S. Tabernaemontani* at banks of rivers and shores of lakes. Adults in May–September. Last instar larvae were found in May, hibernation probably in a larval instar.

Genus *Delphacodes* Fieber, 1866

Delphacodes Fieber, 1866b: 524.

Type-species: *Delphax (Delphacodes) mulsanti* Fieber, 1866, by subsequent designation.

As *Megamelodes*, but styles and appendages of anal tube in male short. Aedeagus with small teeth. Genital scale of female reduced. In Fennoscandia and Denmark two species.

Key to species of *Delphacodes*

1 First antennal segment about as long as apical width. Frons without a light transverse band   48. *venosus* (Germar)
– First antennal segment distinctly longer than maximal width. Lower margin of frons pale   49. *capnodes* (Scott)

48. *Delphacodes venosus* (Germar, 1830)
Text-figs. 378–386.

Delphax venosa Germar, 1830: 57.

Text-figs. 378–386. *Delphacodes venosus* (Germar). – 378: male pygofer from behind; 379: anal tube from behind; 380: anal tube from the left; 381: left genital style from behind; 382: aedeagus, dorsal aspect; 383: aedeagus, ventral aspect; 384: aedeagus from the right; 385: aedeagus from the left; 386: caudal part of female abdomen from below. Scale: 0.2 mm for 386, 0.1 mm for the rest.

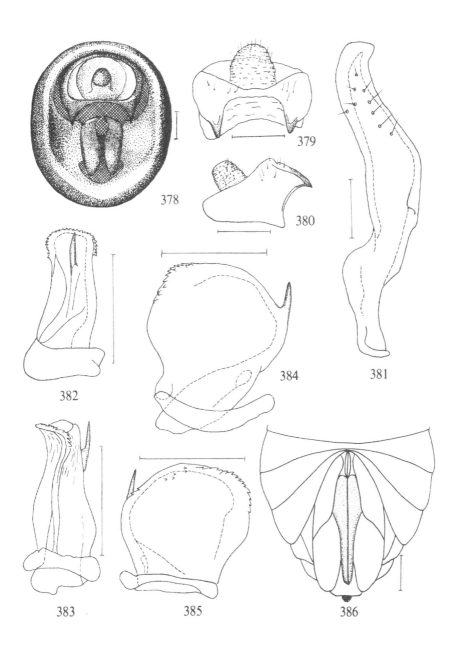

378

379

380

381

382

384

383

385

386

133

Delphax rhypara Flor, 1860: 48.
Liburnia curtula J. Sahlberg, 1871: 450.

Yellowish brown to blackish brown, almost unicolorous. Frons broadest near clypeus, sides in lower half convex. Vertex approximately square, not protruding very much in front of eyes. Fore wings of brachypters with strongly prominent veins, half as long as abdomen or a little longer, about 1.6 times as long as broad, apically rounded. Also in the macropterous form the fuscous veins are strongly raised over the semi-transparent light wing-surface. Genital segment of male (Text-fig. 378) small and short. Anal tube of male (Text-figs. 379, 380) with two short appendages widely apart. Style as in Text-fig. 381, aedeagus as in Text-figs. 382–385. Venter of caudal part of female abdomen as in Text-fig. 386. Length of brachypters 1.4–2.5 mm, of macropters (with wings) 3.0–3.2 mm.

Distribution. Fairly common in Denmark (SJ, EJ, WJ, F, NEZ, B) and Sweden (Sk. – Hls.). – In Norway only found in AK, Drøbak by Warloe (Holgersen, 1946). – Common in Southern and Central East Fennoscandia (Ab, N, Kb, Om, Vib, Kr). – Not in the Pyrenean Peninsula, otherwise widespread in Europe.

Biology. Tyrphobiont. On grasses among *Sphagnum* and other mosses in marshy biotopes, living deep down in the moss cover. Adults in August–June. Hibernation takes place in the adult stage.

49. *Delphacodes capnodes* (Scott, 1870)
Text-figs. 387–395.

Liburnia capnodes Scott, 1870: 69.
Megamelus brevifrons pilosa Haupt, 1935: 134.
Megamelus paludicola Lindberg, 1937: 59.

Shape of head from above and prothorax as in Text-fig. 387. Frons below eyes approximately parallel-sided. Dorsal colour markings of the brachypterous form similar to *Megamelus notula*, but more diffuse. Paler or darker yellowish brown. Terga of thorax and abdomen laterally broadly dark. A narrow pale transverse band present on frons above clypeus. Fore wings of brachypters concolorous, apically rounded with strongly raised tubuliferous veins, 1.7 times as long as broad, 2/3 of length of abdomen. Dorsum of macropters more or less unicolorous, fore wings transparent, brownish with strongly marked veins, the latter with regular series of distinct setigerous tubercles. Male pygofer as in Text-fig. 388, anal tube of male (Text-figs. 389–390) with two short and

Text-figs. 387–395. *Delphacodes capnodes* (Scott). – 387: head and prothorax from above; 388: male pygofer from behind; 389: male anal tube from behind; 390: male anal tube from the left; 391: left genital style; 392: aedeagus, ventral aspect; 393: aedeagus from the right; 394: aedeagus from the left; 395: caudal part of female abdomen from below. Scale: 0.2 mm for 395, 0.1 mm for the rest.

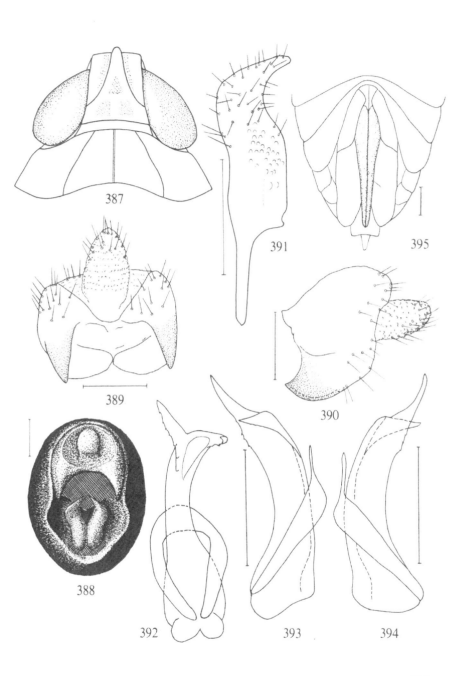

387

391

395

389

390

388

392

393

394

stout appendages situated widely apart. Styles hook-like, together resembling a pair of forceps (Text-fig. 391). Aedeagus as in Text-figs. 392–394; venter of caudal part of female abdomen as in Text-fig. 395. Overall length of macropters 2.8–3.5 mm, of brachypters 1.85–3.5 mm.

Distribution. Sweden: Sk., Bjärred 4.X.1973 (H. Andersson); Sm., Ö. Torsås, Ingelstad 16.VI.1973 (N. Gyllensvärd); Upl., Djursholm, lake Ösbysjön in September and October 1941, 1942, and 1947 (C. H. Lindroth and Ossiannilsson). – So far not recorded from Denmark, Norway, or East Fennoscandia. – England, Netherlands, German D.R. and F.R., Estonia, Poland, Hungary, Yugoslavia.

Biology. Tyrphophilous. Possibly on *Eriophorum* (Strübing, 1956). In *Sphagnum*-tussocks (Schiemenz, 1976). Hibernation in the adult stage (Müller, 1957, Remane, 1958).

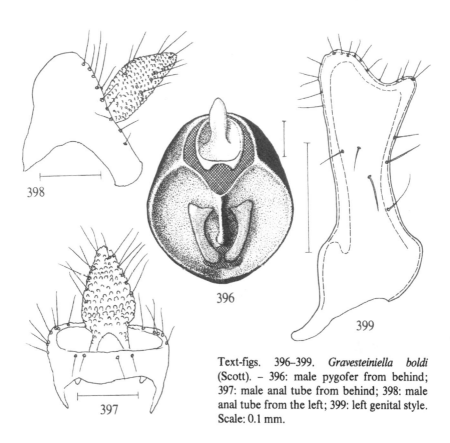

398

396

399

397

Text-figs. 396–399. *Gravesteiniella boldi* (Scott). – 396: male pygofer from behind; 397: male anal tube from behind; 398: male anal tube from the left; 399: left genital style. Scale: 0.1 mm.

Genus *Gravesteiniella* W. Wagner, 1963

Gravesteiniella W. Wagner, 1963: 168.
  Type-species: *Liburnia boldi* Scott, 1870, by original designation.

Frons broadest between lower part of eyes, carinae distinct. Vertex just longer than maximal width, carinae distinct. Lateral carinae of pronotum curving outwards

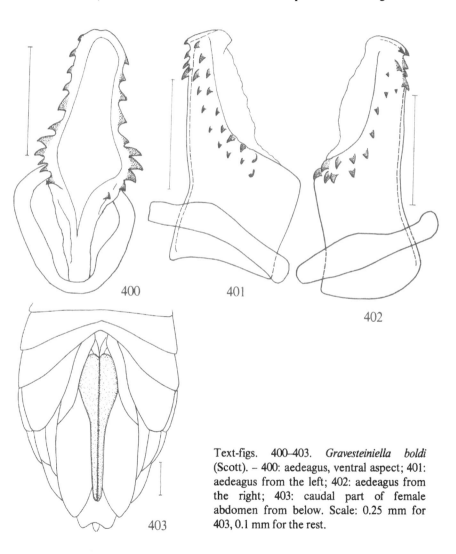

400

401

402

403

Text-figs. 400–403. *Gravesteiniella boldi* (Scott). – 400: aedeagus, ventral aspect; 401: aedeagus from the left; 402: aedeagus from the right; 403: caudal part of female abdomen from below. Scale: 0.25 mm for 403, 0.1 mm for the rest.

posteriorly, not reaching hind margin. Post-tibial calcar with 15–22 marginal teeth, apical tooth small. Fore wing of brachypters about 1.4 times as long as broad, apically rounded. Appendages of anal tube of male very short, blunt. Genital scale of female reduced. One species.

50. *Gravesteiniella boldi* (Scott, 1870)
Text-figs. 396–403.

Liburnia boldi Scott, 1870: 68.

Brownish yellow to dirty yellow. Carinae of head usually pale, margined with black. Notum without a pale median stripe. Abdomen of male black with a median stripe consisting of narrow light spots, lateral and sometimes apical margins of abdominal terga narrowly brownish yellow. Male pygofer black. Abdomen of female dirty yellow, indistinctly dark-mottled, or largely fuscous. Genital phragm of male with a median carina ending below in a ball-like projection (Text-fig. 396). Appendices of anal tube short (Text-figs. 397, 398). Styles as in Text-fig. 399, aedeagus as in Text-figs. 400–402. Ventral aspect of caudal part of female abdomen as in Text-fig. 403. Genital scale absent. Overall length of macropters 4.0–4.6 mm, of brachypters 2.5–3.2 mm.

Distribution. Scarce in Denmark (SJ, EJ, WJ, NWJ, LFM, NEZ, B). – Scarce also in the south of Sweden (Sk., Hall., Gtl., G. Sand.). – Not yet found in Norway. – In East Fennoscandia only in N, Tvärminne (Albrecht, 1977). – England, Scotland, Ireland, Netherlands, German D.R. and F.R., Estonia, Latvia, Poland, Hungary, Slovakia, s. Russia, Cyprus, Altai, Kazakhstan, m. Siberia, Mongolia, Maritime Territory.

Biology. On *Ammophila arenaria* on coastal dunes, adults in June–August.

Genus *Muellerianella* W. Wagner, 1963

Muellerianella W. Wagner, 1963: 168.
Type-species: *Delphax fairmairei* Perris, 1857, by original designation.

Vertex about square. Frons broadest between lower margin of eyes. Lateral margins of frons below eyes almost straight. Median keel of frons forked distinctly below junction to vertex, angle between branches small. Lateral carinae of pronotum curving outwards posteriorly, not reaching hind border. Post-tibial calcar with 15–20 marginal teeth. Fore wings of brachypters apically rounded, 1.6–2.1 times as long as broad. Intermediary specimens with fore wings 2.4–2.5 times as long as broad do exist. Anal tube of male without appendages. Genital scale of female reduced. Larvae whitish with a dark longitudinal stripe on each side of body. Two North-European species; separation of females very difficult.

138

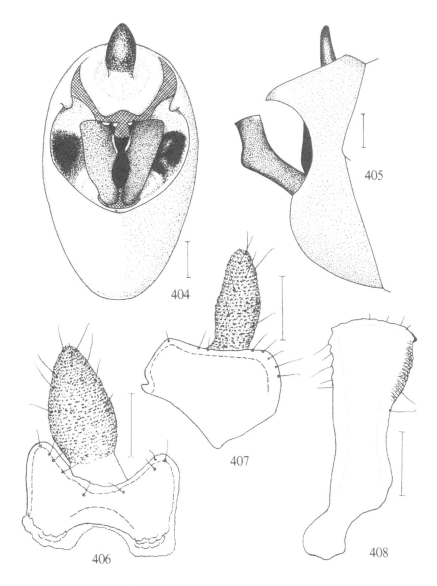

Text-figs. 404–408. *Muellerianella brevipennis* (Boheman). – 404: male pygofer from behind; 405: male pygofer from the right; 406: male anal tube from behind; 407: male anal tube from the left; 408: left genital style. Scale: 0.1 mm.

Key to species of *Muellerianella*

1 Dorsal apical projection of male pygofer sharply pointed, apex directed somewhat downwards (Text-fig. 405). Median carina of frons on junction with vertex often more or less obsolete   51. *brevipennis* (Boheman)
- Dorsal apical projection of pygofer blunt (Text-fig. 415). Median carina of frons sharp throughout   52. *fairmairei* (Perris)

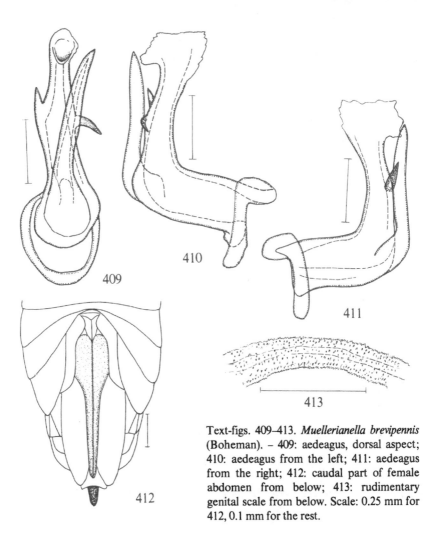

Text-figs. 409–413. *Muellerianella brevipennis* (Boheman). – 409: aedeagus, dorsal aspect; 410: aedeagus from the left; 411: aedeagus from the right; 412: caudal part of female abdomen from below; 413: rudimentary genital scale from below. Scale: 0.25 mm for 412, 0.1 mm for the rest.

51. *Muellerianella brevipennis* (Boheman, 1847)
   Text-figs. 404–413.

Delphax brevipennis Boheman, 1847b: 266.
Delphax bivittata Boheman, 1850: 259.
Delphax hyalinipennis Stål, 1854: 194.

Dirty yellow to brownish yellow. Frons unicolorous or indistinctly mottled brown and yellowish. In dark specimens there is a tendency of light transverse bands appearing between eyes. Ventral border of frons usually lighter. Clypeus often fuscous between keels. First antennal segment apically fuscous. Pro- and mesonotum laterally of side carinae often dark, especially in macropters. Metathorax laterally (on metepisternum) with a fuscous or black spot. Abdomen dorsally often with a broad dark longitudinal band on each side. Fore wings of brachypters yellowish, transparent, covering about 2/3 of abdomen. Fore wings of intermediary form about as long as abdomen, those of macropters much longer, transparent, veins darkened towards apices. Genital styles of males and anal style in both sexes black. Male pygofer as in Text-figs. 404, 405, anal tube of male as in Text-figs. 406, 407, genital style of male as in Text-fig. 408, aedeagus as in Text-figs. 409–411. Ventral aspect of posterior part of female abdomen as in Text-fig. 412, rudiment of genital scale as in Text-fig. 413. Overall length of macropters 3.5–4.9 mm, brachypters 2.5–3.6 mm.

Distribution. Scarce in Denmark, found in SJ, EJ, F, LFM. – Common in the south of Sweden, Sk. – Dlr. – Siebke (1874) recorded *brevipennis* from Norway: AK, Oslo; Holgersen found the species in HEs, Løten. – Common in southern and central East Fennoscandia: Al, Ab – Om, Vib, Kr. – Widespread in Europe.

Biology. According to Drosopoulos (1975, 1977) the host-plant is *Deschampsia flexuosa*. The species is bivoltine in Holland (Drosopoulos, l. c.). Hibernation takes place in the egg stage. In Sweden adults appear in July and August.

52. *Muellerianella fairmairei* (Perris, 1857)
   Text-figs. 414–423.

Delphax fairmairii (sic) Perris, 1857: 170.
Delphax neglecta Flor, 1861: 57.

Resembling *brevipennis*. For differences, see key and text-figures. No tenable morphological differences between females of *brevipennis* and *fairmairei* have been found so far. Male pygofer as in Text-figs. 414, 415, anal tube of male as in Text-figs. 416, 417, genital style as in Text-fig. 418, aedeagus as in Text-figs. 419–421. Venter of caudal part of female abdomen as in Text-fig. 422, rudimentary genital scale as in Text-fig. 423. Overall length of macropters 3.5–4.5 mm, brachypters 2.0–3.0 mm.

Distribution. Scarce in Denmark, found in SJ, EJ, WJ, NWJ, LFM, NEZ, B. – In Sweden less common than *brevipennis*, found in Sk., Bl., Sm., Gtl., Ög., Upl., Vstm. – I

141

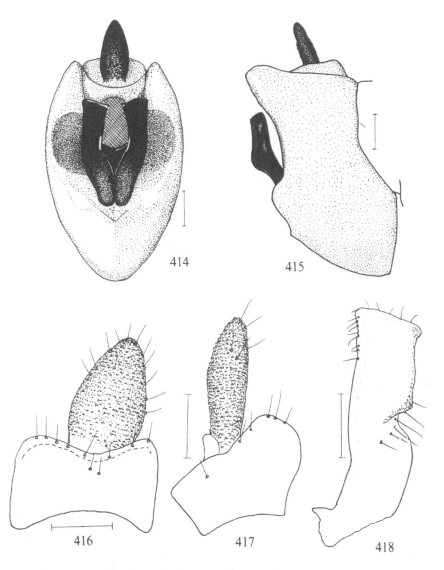

Text-figs. 414–418. *Muellerianella fairmairei* (Perris). – 414: male pygofer from behind; 415: male pygofer from the right; 416: male anal tube from behind; 417: male anal tube from the left; 418: left genital style. Scale: 0.1 mm.

have seen specimens from Ø, VE, VAy, and STi in Norway. – Rare in East Fennoscandia, found in St, Ta, Tb, and Kr. – Widespread in Europe. Also recorded from the Azores, China, Maritime Territory, and Japan, but "the populations from the Azores, Japan and China need to be reexamined" (Drosopoulos, 1977).

Biology. ". . . only in localities where either *Holcus lanatus* or *H. mollis* (food plants) and *Juncus effusus* were growing is close proximity. The latter is used only as an oviposition plant for overwintering eggs". (Drosopoulos, 1977). According to the same

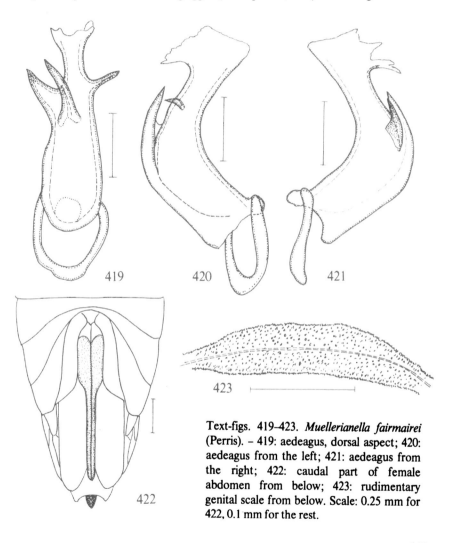

Text-figs. 419–423. *Muellerianella fairmairei* (Perris). – 419: aedeagus, dorsal aspect; 420: aedeagus from the left; 421: aedeagus from the right; 422: caudal part of female abdomen from below; 423: rudimentary genital scale from below. Scale: 0.25 mm for 422, 0.1 mm for the rest.

143

author (1975), *M. fairmairei* is bivoltine in Holland. "The percentage of *M. fairmairei* males varied from 0–25, depending on the area sampled . . ." (Drosopoulos, 1975). This sex-ratio "is due to the coexistence of two female biotypes. One of the biotypes is diploid and bisexual. The other is triploid, most probably of hybrid origin, reproducing gynogenetically, but requiring to be mated with males of the bisexual species in order to give progeny (pseudogamy)" (Drosopoulos, 1977). In Sweden and Norway adults have been found in June–October.

Genus *Muirodelphax* W. Wagner, 1963

Muirodelphax W. Wagner, 1963: 169.
Type-species: *Delphax aubei* Perris, 1857: 170.

Frons broadest between lower part of compound eyes, median carina distinct, obsolete on junction with vertex. Vertex almost pentagonal, just longer than maximal width, fore border obtusely angular or rounded. Lateral carinae of pronotum curving outwards posteriorly, not reaching hind border. Fore wings of brachypters apically oblique with rounded corners, index length: maximal width = 1.4–1.8. Post-tibial calcar with 15–20 marginal teeth, apical tooth missing or very small. Appendages of anal tube in male small or absent. Genital scale of female (in our species) small but distinct. In Europe one species.

53. *Muirodelphax aubei* (Perris, 1857)
Plate-fig. 34, text-figs. 424–433.

Delphax aubei Perris, 1857: 170.
Delphax obsoleta Kirschbaum, 1868: 33.
Liburnia obsoleta J. Sahlberg, 1871, 453.

Head and thorax greyish yellow, dorsum with an indistinctly lighter median longitudinal line. Frons and vertex uniformly greyish yellow, or carinae indistinctly bordered with fuscous. Metathorax laterally partly black or fuscous, legs yellowish with or without longitudinal dark streaks. Fore wings of brachypters covering abdominal segments I–IV, concolorous with notum, semi-transparent, veins often darker. Veins of fore wings in macropters fuscous towards apices. Abdomen in male black with longitudinal rows of wedge-shaped light spots, pygofer entirely or almost entirely black. Abdomen of female dirty yellow, laterally with longitudinal rows of fuscous spots. Pygofer of male as in Text-figs. 424, 425, appendages of anal tube (Text-figs. 426, 427)

Text-figs. 424–433. *Muirodelphax aubei* (Perris). – 424: male pygofer from behind; 425: male pygofer from the right; 426: anal tube from behind; 427: anal tube from the left; 428: left genital style, lateral aspect; 429: aedeagus, dorsal aspect; 430: aedeagus from the left; 431: aedeagus from the right; 432: caudal part of female abdomen from below; 433: genital scale from above. Scale: 0.25 mm for 432, 0.1 mm for the rest.

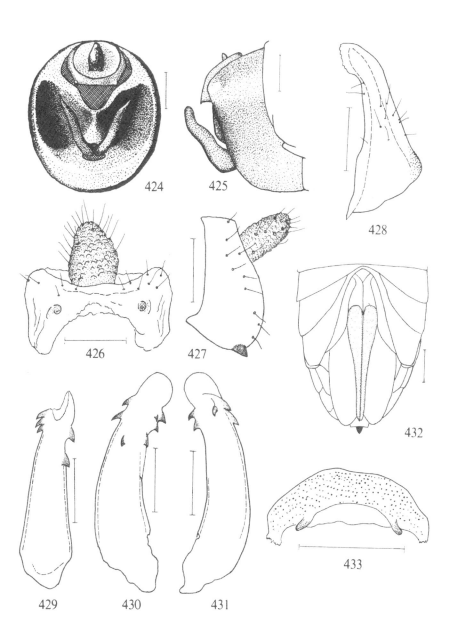

424   425

428

426   427

432

429   430   431   433

very small, placed widely apart. Genital style as in Text-fig. 428, aedeagus as in Text-figs. 429–431. Venter of posterior part of female abdomen and genital scale as in Text-figs. 432 and 433. Overall length of macropters 3.5–4.1 mm, of brachypters 2.1–3.3 mm.

Distribution. Fairly common in Denmark, found in all districts except LFM, SZ, and B. - Fairly common and locally abundant in the south of Sweden: Sk., Hall., Öl., Gtl., Ög. - So far not recorded from Norway. - In East Fennoscandia only found in Al, Lemland (Huldén, 1975). - Widespread in Europe except in the north, also recorded from Tunisia, Anatolia, Georgia, Kazakhstan, Kirghizia, Tadzhikistan, Uzbekistan, and Mongolia.

Biology. On grass in dry biotopes: coastal dunes but also dry meadows, slopes and steppe-like fields like the Alvar of Öland. Hibernation takes place in the larval stage (Müller, 1957, Schiemenz, 1969). In German D.R. there are 2 generations p. a. (Schiemenz, l. c.). In Sweden adults were found in July and August.

Genus *Acanthodelphax* Le Quesne, 1964

Acanthodelphax Le Quesne, 1964: 57.
Type-species: *Delphax denticauda* Boheman, 1847, by original designation.

Vertex distinctly broader than median length, fore border rounded. Frons broad, broadest between lower margins of compound eyes, sides arched, convex. Median keel of frons distinct, obsolete on junction with vertex. Lateral carinae of pronotum curving outwards posteriorly, not reaching hind border. Fore wings of brachypters apically truncate, 1.2–1.45 times as long as broad. Marginal teeth of post-tibial calcar few in number. Male pygofer ventrally with sharp upturned median process. Anal tube of male without paired appendages. In Northern Europe one species.

54. *Acanthodelphax denticauda* (Boheman, 1847)
Text-figs. 434–442.

Delphax denticauda Boheman, 1847a: 64.
Liburnia oxyura Haupt, 1935: 145.

Frons and vertex usually entirely yellowish, or carinae of frons indistinctly bordered with fuscous. Pro- and mesonotum of male dirty yellowish, sides and venter of thorax

---

Text-figs. 434–442. *Acanthodelphax denticauda* (Boheman). – 434: male pygofer from behind; 435: male pygofer from the right; 436: male anal tube from the right; 437: left genital style; 438: aedeagus, dorsal aspect; 439: aedeagus from the right; 440: caudal part of female abdomen from below; 441: genital scale from below; 442: median scleroite between basal parts of ovipositor from above. Scale: 0.25 mm for 440, 0.1 mm for the rest.

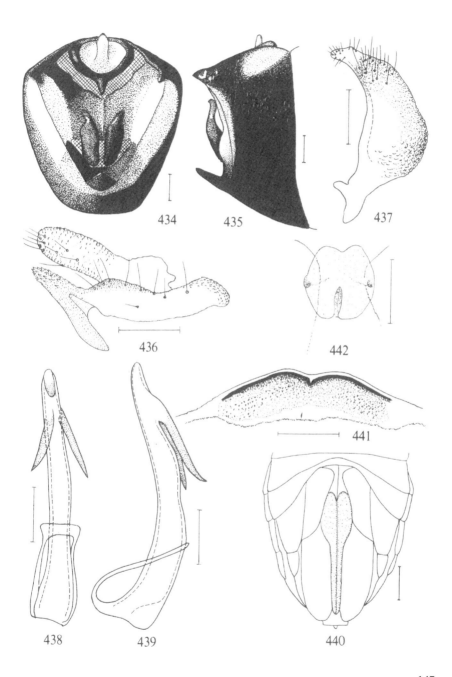

434 435 437

436 442

438 439 440

441

147

largely black. Fore wings of brachypters not reaching hind border of 4th abdominal segment. Abdomen of male black with an indistinct median stripe and hind borders of posterior segments light. Male pygofer large, black, with hind border partly light (Text-figs. 434, 435). Female usually unicolorous, light yellow. Male anal tube (Text-fig. 436) with an unpaired median process. Genital style of male (Text-fig. 437) short, stout, distally suddenly narrower, apex blunt. Aedeagus as in Text-figs. 438, 439. Lateral lobes of female broad (Text-fig. 440). Genital scale rudimentary (Text-fig. 441). Caudally of the genital scale, between basal parts of the ovipositor, is an unpaired median sclerite shaped as in Text-fig. 442. Overall length of macropters 3–4 mm, of brachypters 2–3 mm.

Distribution. Scarce in Denmark, found in EJ, LFM, NWZ, and NEZ. – Not rare, locally abundant, in Sweden (Öl., Ög., Upl., Dlr., Ång., P. Lpm.). – Siebke (1874) recorded *denticauda* from Norway: AK, Oslo. – Scarce and sporadic in East Fennoscandia but found in most districts from Ab to LkE (also Vib and Kr). – Not found in the Mediterranean countries, otherwise widespread in Europe.

Biology. In moist meadows and forest glades. According to Raatikainen & Vasarainen (1976) also in leys, field margins and cereal fields. Linnavuori (1952) suggested that *A. denticauda* probably lives on *Calamagrostis lanceolata*. Drosopoulos (1977) reared it in the laboratory on *Deschampsia caespitosa*. In Sweden adults were found in May, June, and July.

Genus *Tyrphodelphax* Vilbaste, 1968

Tyrphodelphax Vilbaste, 1968: 72.
Type-species: *Delphax distincta* Flor, 1861, by original designation.

Vertex as long as broad or slightly longer, anterior margin convex. Frons widest near middle, sides weakly convex. Lateral pronotal carinae not reaching hind margin of pronotum, posteriorly curving outwards. Post-tibial calcar without teeth along lower margin, only apical tooth present. Appendages of anal tube in male moderately long. Genital scale of female distinct, partly strongly sclerotized. In Europe two tyrphophilous species.

Text-figs. 443–453. *Tyrphodelphax distinctus* (Flor). – 443: male pygofer from behind; 444: male pygofer from the right; 445: male anal tube from behind; 446: male anal tube from the left; 447: left style; 448: aedeagus, ventral aspect; 449: aedeagus from the right; 450: aedeagus from the left; 451: apex of aedeagus, dorsal aspect; 452: caudal part of female abdomen from below; 453: genital scale from above. Scale: 0.25 mm for 452, 0.1 mm for the rest.

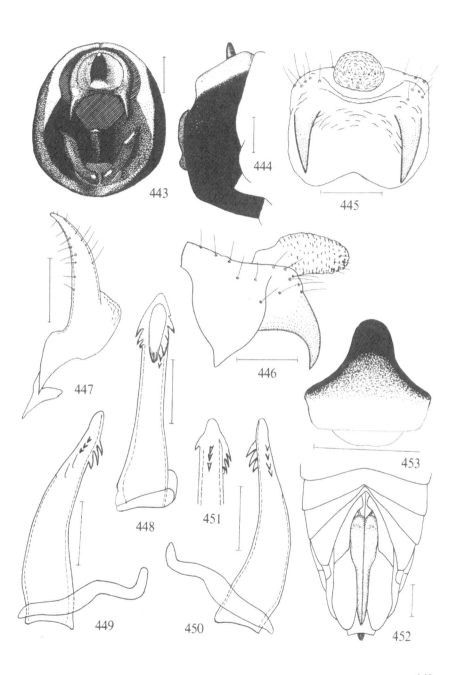

443

444

445

446

447

448

449

450

451

452

453

149

55. *Tyrphodelphax distinctus* (Flor, 1861)
Text-figs. 443–453.

Delphax distincta Flor, 1861: 68.
Liburnia albocarinata forma *brachyptera* J. Sahlberg, 1871: 426.
Calligypona albocarinata Ossiannilsson, 1946c: 57 (p. p.).

Frons black between ivory-white carinae. Vertex between posterior carinae orange. Pronotum and scutellum yellowish, sometimes with diffuse fuscous markings, carinae white, median carina of scutellum bordered with white. Fore wings of brachypters light yellow or brownish with white margins, covering the four basal abdominal segments. Legs sordid yellow with fuscous longitudinal streaks. Male abdomen black, dorsum medially with a series of segmental triangular light spots, dorsum of abdominal segment VIII and dorsal half of hind border of pygofer whitish. Abdomen of female sordid brownish yellow with diffuse, more or less extended, dark markings. Male pygofer as in Text-figs. 443, 444, appendices of male anal tube (Text-figs. 445, 446) moderately long, widely apart from each other, genital styles (Text-fig. 447) slightly curved, acute-pointed, aedeagus as in Text-figs. 448–451, venter of caudal part of female abdomen as in Text-fig. 452, genital scale as in Text-fig. 453. Length of brachypters 2–2.9 mm. I have not seen macropterous specimens of this species.

Distribution. Scarce in Denmark (NEZ, B). – Fairly common in Sweden, Sk. – P. Lpm. – Norway: HEs, HEn, Ri, HOi, NTi. – Comparatively common in central and northern parts of East Fennoscandia; Kb, Ok, ObN, LkE, Li. – England, Scotland, Ireland, Estonia, Latvia, Poland, German D.R. and F.R., Switzerland, Austria, Bohemia.

Biology. On *Eriophorum vaginatum*. "*C. distincta* legt die Eier von der Blattunterseite in das fleischige *Eriophorum*-Blatt" (Strübing, 1956). Hibernation takes place in the larval stage (Strübing in Müller, 1957). In Sweden adults have been found in May–August.

Text-figs. 454–463. *Tyrphodelphax albocarinatus* (Stål). – 454: male pygofer from behind; 455: male pygofer from the right; 456: male anal tube from behind; 457: male anal tube from the left; 458: left genital style; 459: aedeagus, ventral aspect; 460: aedeagus from the left; 461: aedeagus from the right; 462: caudal part of female abdomen from below; 463: genital scale from above. Scale: 0.25 mm for 462, 0.1 mm for the rest.

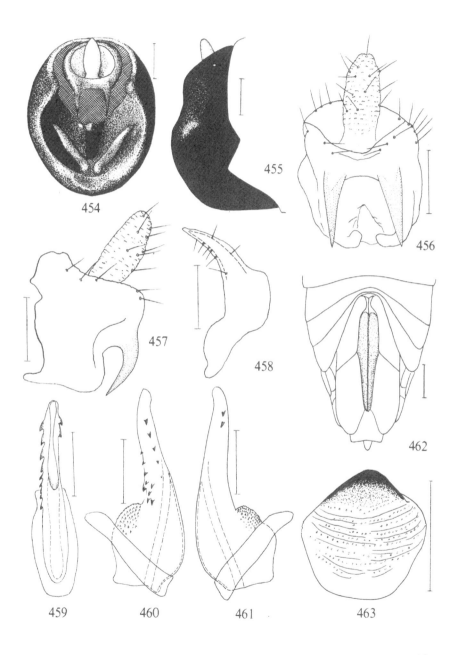

454

455

456

457

458

462

459   460   461   463

151

56. **Tyrphodelphax albocarinatus** (Stål, 1858)
Text-figs. 454-463.

Delphax albocarinata Stål, 1858: 357.
Liburnia albocarinata forma *intermedia* J. Sahlberg, 1871: 426.
Calligypona albocarinata Ossiannilsson, 1946c; 57 (p. p.).

Very like the preceding species, differing by characters given in the key and by the structure of male and female genitalia. Male pygofer as in Text-figs. 454, 455, anal tube of male as in Text-figs. 456, 457, genital style as in Text-fig. 458, aedeagus as in Text-figs. 459-461, venter of female abdomen as in Text-fig. 462, genital scale as in Text-fig. 463. Length of brachypters 2.1-2.9 mm, macropters 4 mm.

Distribution. So far not found in Denmark, nor in Norway. – Rare in Sweden. Described on material from Ång., Hällingsås and Forse (Stål). Later collected in Sk., Svalöf (Ossiannilsson), Sm., Ö. Korsberga (D. Gaunitz), Vrå (Ossiannilsson), Ög., Vårdsberg (Ossiannilsson), Jmt., Sunne, Svedje, Bleksjön (Ossiannilsson), Vb. Skellefteå (Ossiannilsson), vicinity of Hällnäs (D. Westerberg). – East Fennoscandia: distribution imperfectly investigated, recorded from Al, Eckerö (Reuter); Ab, Pargas (Reuter), Karislojo (J. Sahlberg), Sammatti (J. Sahlberg). Also found in Kb, Ok, and ObN. – Estonia, Poland, German D.R. and F.R., Austria, Bohemia, n. Russia.

Biology. On *Eriophorum*. Hibernation in the larval stage (Schiemenz, 1976). In Sweden adults appear in June–August.

Genus *Dicranotropis* Fieber, 1866

Dicranotropis Fieber, 1866b: 530.
Type-species: *Delphax hamata* Boheman, 1847, by subsequent designation.

Median carina of frons forked considerably below junction with vertex (Text-fig. 464), or frons with two median carinae. Vertex almost square, rarely distinctly longer than wide, anteriorly with four parallel carinae. Pro- and mesonotum each with three carinae, lateral carinae of pronotum strongly curved, not reaching hind border. Teeth of post-tibial calcar small and few in number. In Denmark and Fennoscandia one species.

57. **Dicranotropis hamata** (Boheman, 1847)
Plate-fig. 18, text-figs. 464-475.

Delphax hamata Boheman, 1847a: 45.

Fore wings of brachypters apically rounded, reaching to or somewhat apically of hind border of 4th abdominal tergum. Brownish yellow. Carinae of head ivory-white, bordered with black, vertex, pronotum, and mesonotum with an ivory-white median

152

longitudinal band. In the male (Plate-fig. 18), the venter of thorax, hind femora, and abdomen are black, abdomen with a median longitudinal series of light spots and a few light patches near basis, dorsum of segment VIII and pygofer orange or yellowish. Female brownish yellow or dirty yellow with more or less extended diffuse fuscous patches on venter of thorax and dorsum of abdomen. Fore wings semi-transparent, in

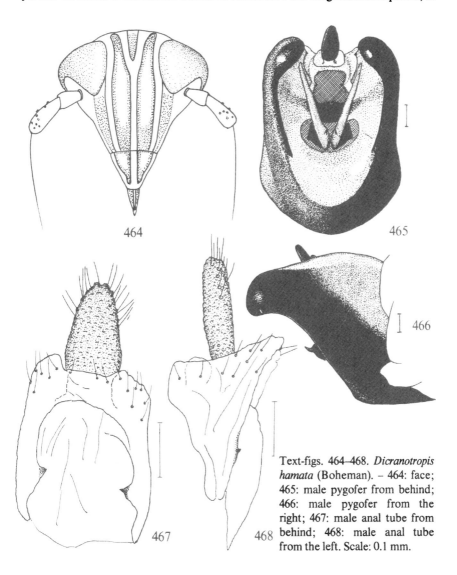

Text-figs. 464–468. *Dicranotropis hamata* (Boheman). – 464: face; 465: male pygofer from behind; 466: male pygofer from the right; 467: male anal tube from behind; 468: male anal tube from the left. Scale: 0.1 mm.

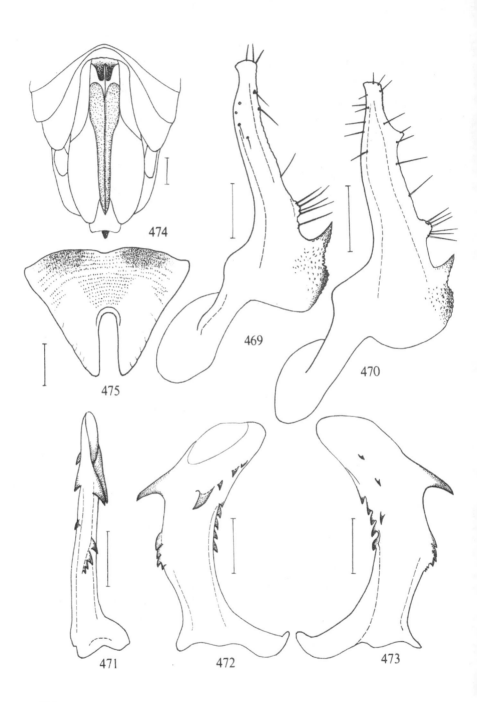

474

469

470

475

471

472

473

154

macropters with darker veins, in both sexes (brachypters and macropters) with a blackish spot or longitudinal streak in apex of clavus. Sometimes this spot is indistinct in brachypters. Pygofer of male (Text-figs. 465, 466) large, in dorsal aspect deeply emarginate, genital phragm yellowish, anal style black, anal tube (Text-figs. 467, 468) with two small tooth-like appendages. Genital styles of male (Text-figs. 469, 470) long, somewhat varying in structure. Aedeagus as in Text-figs. 471–473. Ventral aspect of posterior part of female abdomen as in Text-fig. 474, anal style of female black, genital scale (Text-fig. 475) large, caudally deeply emarginate. Overall length of macropters 4.25–4.9 mm, of brachypters 2.85–4.0 mm.

Distribution. Common and widespread in Denmark, also in Sweden (Sk. – Vb.). – In Norway not rare, often abundant, found in AK, Os, Bø, TEi, AAy, HOi, SFi, and NTi. – East Fennoscandia: very common in South and Central Finland. "In southern Finland up to a latitude of about 63°N" (Raatikainen & Vasarainen, 1964). – Widespread in Europe, also found in Algeria, Tunisia, Kazakhstan, Altai, m. Siberia, Mongolia.

Biology. *Dicranotropis hamata* is common and often abundant in grass meadows, woods and cultivated fields. Its biology has been studied in England by Hassan (1939) and in Finland by Raatikainen & Vasarainen (1964). The species has been reared on oats, wheat, timothy, *Deschampsia caespitosa*, and *Agrostis tenuis* (Raatikainen & Vasarainen, l. c.), and feeds on many other grasses as *Holcus lanatus, Elytrigia repens, Arrhenatherum elatius, Alopecurus pratensis*, and *Lolium perenne* (Hassan, l. c.). Eggs are deposited in groups in stems and leaves of grasses, the number of eggs per group ranging from 1 to 41. There is only one generation per annum in Finland, and hibernation takes place in nymphal stages II–IV. In Finland and also in Sweden brachypters are more numerous than macropters. For data on natural enemies of this species the reader is referred to Raatikainen & Vasarainen (l. c.). In Sweden adults have been found from end of May to beginning of October.

Economic importance. *Dicranotropis hamata* is a vector of the virus causing the oat dwarf disease of cereals (Lindsten, 1961), and also of the oat sterile-dwarf virus (Ikäheimo & Raatikainen, 1963), and of the cereal tillering disease (Lindsten & Gerhardsen, 1971).

Text-figs. 469–475. *Dicranotropis hamata* (Boheman). – 469: right genital style from inside (Swedish specimen); 470: same (specimen from Stettin); 471: aedeagus, dorsal aspect; 472: aedeagus from the left; 473: aedeagus from the right; 474: caudal part of female abdomen from below; 475: genital scale from below. Scale: 0.25 mm for 474, 0.1 mm for the rest.

155

Genus *Florodelphax* Vilbaste, 1968

Florodelphax Vilbaste, 1968: 70.
Type-species: *Delphax paryphasma* Flor, 1861, by original designation.

Vertex not longer than wide. Frons 1.7–1.8 times as long as broad. Sides of frons distinctly convex. Keels of pronotum curved outwards, not reaching hind border. Fore wings of brachypters apically truncate, black (♂) or brownish (♀), apical margin white. Post-tibial calcar with 10–18 marginal teeth, apical tooth small. Pygofer of male without a lateral incision. Appendages of anal tube of male long, pointed. Genital styles moderately long, diverging. In Europe two species.

Key to species of *Florodelphax*

1. Carinae of frons obsolescent on transition with vertex. Appendages of male anal tube slightly diverging, basally almost contiguous (Text-figs. 476, 478). Aedeagus as in Text-figs. 481–483. Incision near basis of lateral lobe of female rounded, not angular (Text-fig. 484)   58. *paryphasma* (Flor)
   – Carinae of frons and vertex not obsolescent. Appendages of anal tube not diverging, basally well apart from each other (Text-figs. 486, 488). Aedeagus as in Text-figs. 491–493. Incision near basis of lateral lobe of female angular (Text-fig. 494)   59. *leptosoma* (Flor)

58. *Florodelphax paryphasma* (Flor, 1861)
Text-figs. 476–485.

Delphax paryphasma Flor, 1861: 75.
Liburnia niveimarginata Scott, 1870: 71.
Delphax lucticolor J. Sahlberg, 1868: 189.
Calligypona leptosoma Ossiannilsson, 1946c: 62, nec Flor.

In both sexes reminding of *Criomorphus albomarginatus*. Head brownish yellow. Frons 1.7–1.8 times as long as broad, broadest between lower margins of eyes. Thorax in brachypters brownish yellow, venter in male partly black. Mesonotum and caudal part of pronotum whitish. Fore wings of brachypters apically truncate, 1.4–1.5 times as long as broad. Abdomen of male black or dark brown, small lateral spots, hind margins of terga VII and VIII and of pygofer whitish. Anal tube and anal style black. Coxae and

---

Text-figs. 476–485. *Florodelphax paryphasma* (Flor). – 476: male pygofer from behind; 477: male pygofer from the right; 478: male anal tube from behind; 479: male anal tube from the left; 480: left genital style; 481: aedeagus, ventral aspect; 482: aedeagus from the right; 483: aedeagus from the left; 484: caudal part of female abdomen from below; 485: genital scale from below. Scale: 0.25 mm for 484, 0.1 mm for the rest.

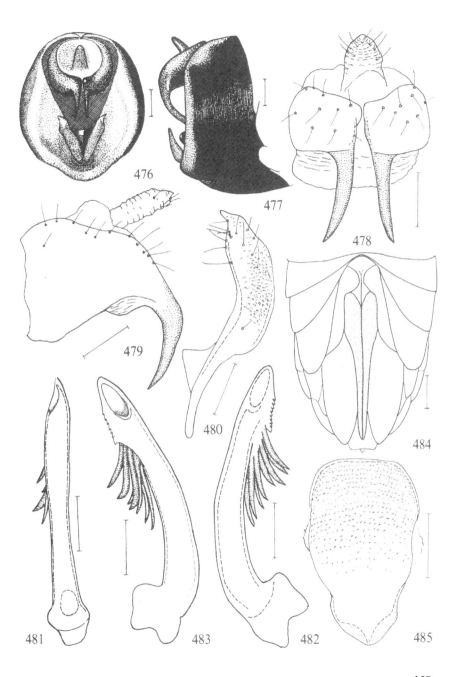

476

477

478

479

480

484

481

483

482

485

femora ∓ fuscous. Brachypterous female brownish yellow to brown or dark brown, dark mottled. Macropters largely dark brown, hind border of pronotum broadly whitish. Male pygofer as in Text-figs. 476, 477, anal tube of male as in Text-figs. 478, 479, genital styles as in Text-fig. 480, aedeagus as in Text-figs. 481–483. Venter of posterior part of female abdomen as in Text-fig. 484, genital scale as in Text-fig. 485. Overall length of brachypters 1.8–3.5 mm, of macropters 3.4–3.9 mm.

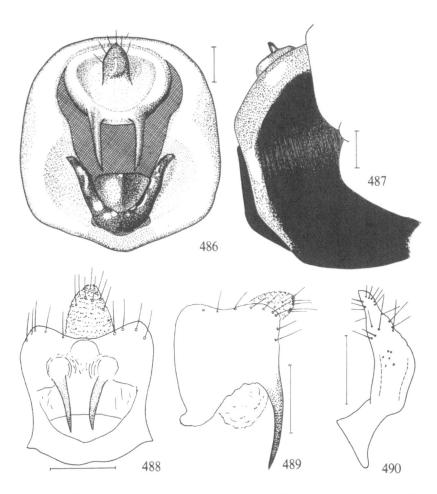

Text-figs. 486–490. *Florodelphax leptosoma* (Flor). – 486: male pygofer from behind; 487: male pygofer from the right; 488: male anal tube from behind; 489: male anal tube from the left; 490: left genital style. Scale: 0.1 mm.

Distribution. So far not found in Denmark, nor in Norway. – Not rare in Sweden, found in Sm., Öl., Gtl., Ög., Sdm., and Upl. – Rare in East Fennoscandia: Al, Eckerö (Håkan Lindberg), Lemland (Hellén); N, Strömfors (Öblom); Ta, Ruovesi (J. Sahlberg); Kr, Keksholm and Petrosavodsk (J. Sahlberg).

Biology. On wet meadows, lake-shores etc., adults in May–August.

59. *Florodelphax leptosoma* (Flor, 1861)
Text-figs. 486–494.

Delphax leptosoma Flor, 1861: 76.
Delphax albofimbriata Fieber, 1866: 534.

Very like the preceding species, differing by carinae of frons and vertex being quite distinct, and by the structure of male and female genitalia. Frons and vertex of male between carinae brown or black. Pronotum of brachypterous male dirty white, anteriorly often dark, scutellum and abdominal tergum shining black, fore wings also black, hind margin white, base sometimes whitish. Fore wings apically truncate, about 1 1/2 times as long as broad. Pygofer black, margined with white. Frons of female between carinae brownish, more or less mottled. Vertex, pronotum, scutellum, and fore wings of brachypterous female lighter or darker brownish yellow, hind margin of fore wing whitish. Scutellum of macropters largely black or black-brown. Male pygofer as in

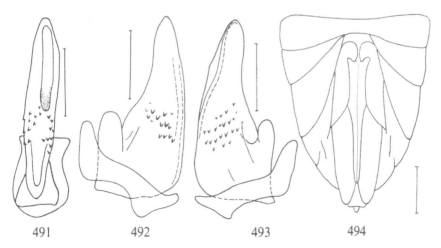

491   492   493   494

Text-figs. 491–494. *Florodelphax leptosoma* (Flor). – 491: aedeagus, ventral aspect; 492: aedeagus from the left; 493: aedeagus from the right; 494: caudal part of female abdomen from below. Scale: 0.25 mm for 494, 0.1 mm for the rest. (494 after Vilbaste, 1971).

159

Text-figs. 486, 487, anal tube of male as in Text-figs. 488, 489, genital style as in Text-fig. 490, aedeagus as in Text-figs. 491–493, ventral aspect of posterior part of female abdomen as in Text-fig. 494. Length of brachypters 2.0–2.7 mm, of macropters with wings 3.5–4.1 mm.

Distribution. Scarce in Denmark, found in NWJ, NEZ, and B. – In Sweden only found in Sk., Barkåkra, Magnarp 8.VII.1970 (N. Gyllensvärd leg.). – Not in Norway, nor in East Fennoscandia. – Widespread in Europe, also in Anatolia.

Biology. On wet meadows, in the "Cariceto canescentis – Agrostidetum caninae-association" (Marchand, 1953).

Genus *Kosswigianella* W. Wagner, 1963

Kosswigianella W. Wagner, 1963: 169.
Type-species: *Delphax exigua* Boheman, 1847, by original designation.

Vertex about as long as broad, anteriorly more or less rounded. Frons 1.7–1.8 times as long as broad, broadest between lower part of eyes, sides rather evenly curved. Median carina of frons more or less obsolescent on transition with vertex. Lateral carinae of pronotum posteriorly curved outwards, not reaching hind border. Fore wings of brachypters apically truncate, 1.2–1.3 times as long as broad. Marginal teeth of post-tibial calcar few in number (9–12), more sparsely arranged towards apex, apical tooth absent or very small. Processes of anal tube of male small, apices directed towards each other. Genital styles each with a basal pointed process directed backwards. Median margin of lateral lobes of female without an incision. Genital scale of female rudimentary. In Europe one species.

60. *Kosswigianella exigua* (Boheman, 1847)
    Text-figs. 495–504.

Delphax exigua Boheman, 1847a: 65.

Dorsum of head and thorax dirty yellow without a lighter median band. Venter dark spotted. Frons entirely light, or carinae in their lower part indistinctly bordered with fuscous. Clypeus largely fuscous with light carinae. Fore wings of brachypters semi-transparent, dirty yellow, reaching a little beyond hind border of third abdominal

Text-figs. 495–504. *Kosswigianella exigua* (Boheman). – 495: male pygofer from behind; 496: male pygofer from the right; 497: male anal tube from behind; 498: male anal tube from the left; 499: left genital style from the left; 500: aedeagus, dorsal aspect; 501: aedeagus from the left; 502: aedeagus from the right; 503: caudal part of female abdomen from below; 504: rudimentary genital scale from below. Scale: for 503 0.25 mm, for the rest 0.1 mm.

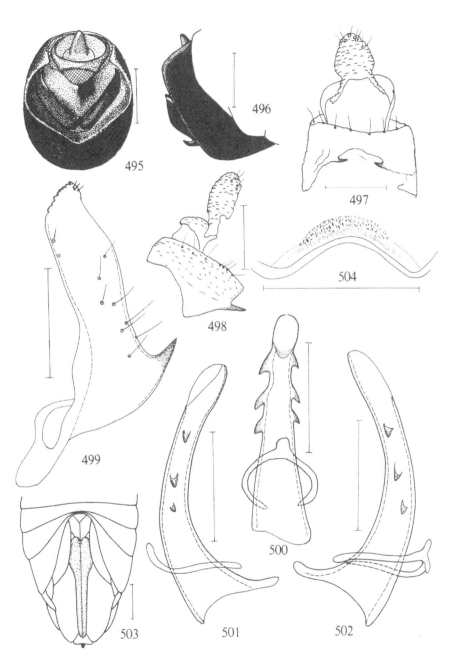

495

496

497

498

504

499

500

501

502

503

161

tergum. Abdomen of male black with a median longitudinal row of narrow light spots, pygofer (Text-figs. 495, 496) small, black, hind border above indistinctly lighter. Abdomen of brachypterous female brownish yellow or dirty with rows of blackish spots and often a lighter median line. Scutellum of macropters often black or fuscous laterally of side keels and with two dark spots anteriorly between keels. Anal tube of male as in Text-figs. 497 and 498, genital style (Text-fig. 499) basally with a backwards directed pointed process. Aedeagus as in Text-figs. 500–502, venter of posterior part of female abdomen as in Text-fig. 503, genital scale rudimentary (Text-fig. 504). Overall length of brachypters 1.6–2.7 mm, of macropters 2.9–3.4 mm.

Distribution. Fairly common in Denmark: SJ, EJ, WJ, LFM, SZ. – Common in the south of Sweden up to Upl. and Vstm. – Not found in Norway. – Rare in East Fennoscandia, found in Al, Eckerö and Sund; Ab, Pargas; Vib, Kivinebb; Kr, Syväri. – Widespread in Europe, present also in Tunisia and Japan.

Biology. On dry meadows, often abundant. According to Kuntze (1937) on "Binnen-dünen, Sandfelder, besonnte Hänge, Waldlichtungen und Wiesen". A member of the "Corynephoretum agrostidetosum aridae" (Marchand, 1953). Eurytopic in xerophilous (and mesophilous) biotopes (Schiemenz, 1969). Hibernation takes place in the larval stage (Kuntze, 1937, Müller, 1957, Remane, 1958, Schiemenz, 1969). In Sweden adults have been found in April–August. Macropters are rare.

Genus *Struebingianella* W. Wagner, 1963

Struebingianella W. Wagner, 1963: 169.
Type-species: *Delphax lugubrina* Boheman, 1847, by original designation.

Vertex just shorter than broad, fore border rounded. Frons 1.8–2.1 times as long as broad, broadest on middle, sides moderately convex. Median carina of frons obsolescent on transition with vertex. Lateral carinae of pronotum posteriorly curved outwards, not reaching hind border. Post-tibial calcar long and slender, with 15–22 marginal teeth, apical tooth small or absent. Fore wings of brachypters apically rounded. Pygofer of male without a lateral incision. Genital styles of male diverging. Appendages of male anal tube moderately long, directed towards venter. In Northern Europe two species.

Key to species of *Struebingianella*

1 Frons and clypeus black or fuscous between light carinae. Male abdomen largely black. Fore wings of brachypters transparent, dirty light yellow. Saw-case of female black or fuscous   62. *litoralis* (Reuter)
– Frons and clypeus concolorous with carinae, light. Male abdomen above partly light. Fore wings of brachypterous male black, basis and margin light, of female transparent, light. Saw-case of female light   61. *lugubrina* (Boheman)

162

61. Struebingianella lugubrina (Boheman, 1847)
Text-figs. 505–514.

Delphax lugubrina Boheman, 1847b: 266.

Side margins of frons weakly and evenly convex. Fore wings of brachypters 1.6–1.7

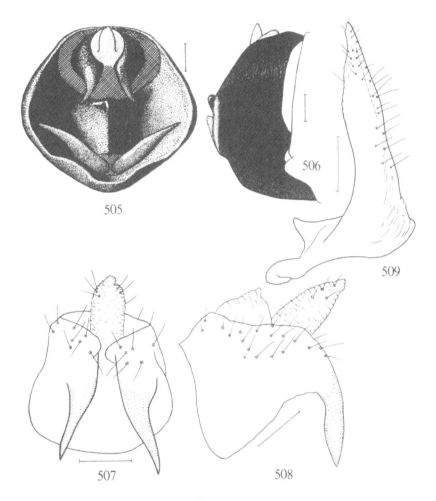

Text-figs. 505–509. *Struebingianella lugubrina* (Boheman). – 505: male pygofer from behind; 506: male pygofer from the right; 507: male anal tube from behind; 508: male anal tube from the left; 509: left genital style. Scale: 0.1 mm.

163

times as long as broad, covering about half abdomen. In brachypterous male, frons, clypeus, vertex, and notum yellowish, abdomen and venter of thorax partly black or fuscous, fore wings brownish black with basis and margin yellowish, hind margins of abdominal terga yellow. Usually, terga of posterior abdominal segments are largely yellow. Pygofer black, anal style light. Coxae black, femora and tibiae yellowish. I have not seen macropterous males of *S. lugubrina*. Brachypterous female entirely yellow or

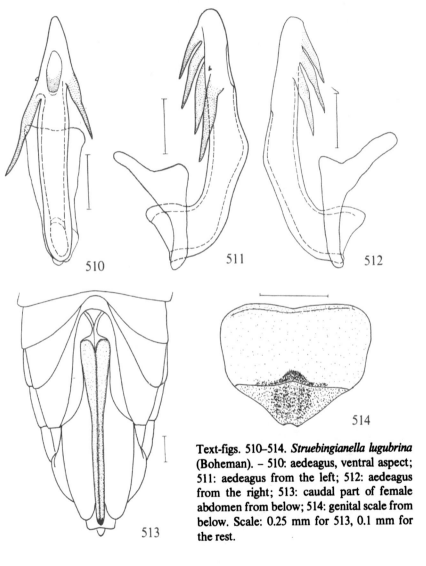

Text-figs. 510–514. *Struebingianella lugubrina* (Boheman). – 510: aedeagus, ventral aspect; 511: aedeagus from the left; 512: aedeagus from the right; 513: caudal part of female abdomen from below; 514: genital scale from below. Scale: 0.25 mm for 513, 0.1 mm for the rest.

brownish yellow, or with a dark longitudinal band on each side of abdominal tergum, and with dark spots on venter. Frons sometimes dark mottled. Macropterous female yellowish, dark spotted on venter and with basal abdominal terga largely fuscous. Male pygofer as in Text-figs. 505, 506, anal tube of male as in Text-figs. 507, 508, genital style as in Text-fig. 509, aedeagus as in Text-figs. 510–512 (the number of appendages of the aedeagus is varying). Venter of posterior part of female abdomen as in Text-fig. 513, genital scale as in Text-fig. 514. Length of brachypterous male 2.5–3.1 mm, of brachypterous female 3.9–4.5 mm, of macropterous male (according to Vilbaste, 1971) 5.05 mm, of macropterous female (with wings) 5.0–5.5 mm.

Distribution. Fairly common in Denmark, found in EJ, F, LFM, SJ, and NEZ. – Sweden not rare, locally abundant, but so far only found in Sk., Bl., Sm., Öl., Gtl., Ög., Vg., Sdm., and Dlr. – Not found in Norway. – Fairly rare in East Fennoscandia, recorded from Al, Ab, N, Ka, Ta, Vib, and Kr. – Widespread in Europe.

Biology. In wet biotopes. In "Flachmooren, Uferzone, Auerwäldern und Erlenbrüchen" (Kuntze, 1937). On Carex and Phragmites (Linnavuori, 1969). I found S. lugubrina with Glyceria maxima, the host-plant according to Müller (1951). In shore meadows and even spruce and birch swamps (Raatikainen & Vasarainen, 1976). Hibernation takes place in larval stages (Müller, 1957). In Sweden adults were found in May–September.

62. **Struebingianella litoralis** (Reuter, 1880)
Text-figs. 515–523.

Liburnia litoralis Reuter, 1880: 198.

Vertex light brown, anteriorly darker. Frons and clypeus black or dark-mottled between yellow-brown carinae. Side margins of frons rather convex. Apex of 1st and basis of 2nd antennal segment often black. Pro-, meso-, and metanotum in brachypters sordid yellow or light brown, fore wings transparent, dirty yellowish, 1.7–1.8 times as long as broad, apically rounded. Abdomen in males largely black, in females light brown with darker markings. I have not seen macropterous specimens. According to Le Quesne (1960), scutellum is blackish in macropterous males, vertex, pronotum and scutellum being largely black-brown in macropterous females. Fore wings of macropterous males are said to be hyaline, almost colourless, those of macropterous females light brownish, veins darker. Pygofer of male as in Text-figs. 515, 516, anal tube (Text-figs. 517, 518) with appendages widely apart, genital style as in Text-fig. 519, aedeagus as in Text-figs. 520–522. Ventral aspect of posterior part of female abdomen as in Text-fig. 523. Length of brachypters 2.1–3.2 mm, of macropters (with wings) 3.5–4.3 mm.

Distribution. Very rare, so far only recorded from Finland and Scotland. Finland: Ab, Pargas, Kapellstrand (Reuter); Oa, Korsholm 13.VI., 28.VI.1940, Kvevlax 14.VI.-1940 (Håkan Lindberg).

165

Biology. Reuter found the species at a sea-shore on *Heleocharis* and *Phragmites*. In Scotland it was taken "on sedges round a small loch in a deep hollow on the moors at Aviemore, Inverness-shire" (Le Quesne, 1960a). Adults in June and July (Le Quesne, 1960b).

Genus *Xanthodelphax* W. Wagner, 1963

Xanthodelphax W. Wagner, 1963: 169.
Type-species: *Delphax flaveola* Flor, 1861, by original designation.

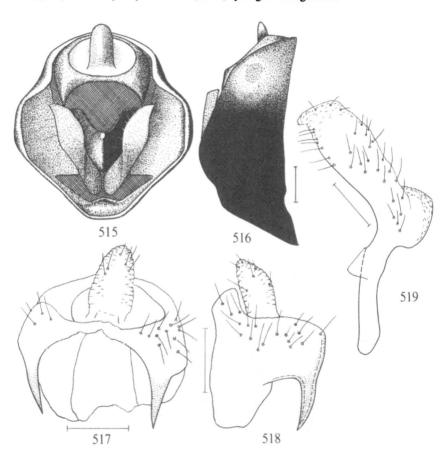

Text-figs. 515–519. *Struebingianella litoralis* (Reuter). – 515: male pygofer from behind; 516: male pygofer from the right; 517: male anal tube from behind; 518: male anal tube from the left; 519: left genital style. Scale: 0.1 mm.

166

Body of brachypters entirely light yellow or light orange. Vertex just shorter than broad. Frons 1.7–1.9 times as long as broad, broadest between lower part of eyes, sides distinctly convex. Lateral carinae of pronotum curved outwards, not reaching hind border. Fore wings of brachypters apically rounded or truncate with rounded corners, 1.5–1.8 times as long as broad. Post-tibial calcar short, number of marginal teeth small (10–11), apical tooth very small. Appendages of anal tube (in our species) short, triangular as seen from behind. Genital styles forceps-like. In Europe three species, two of them present in Denmark and Fennoscandia.

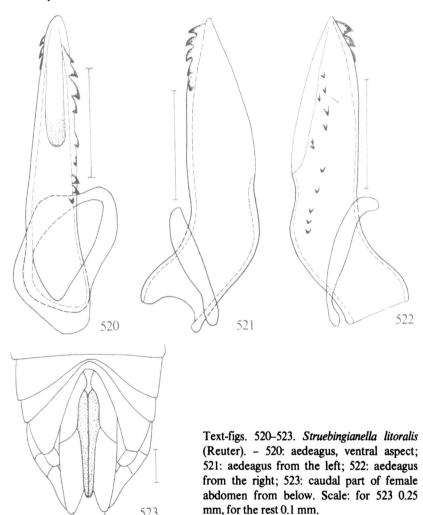

520

521

522

523

Text-figs. 520–523. *Struebingianella litoralis* (Reuter). – 520: aedeagus, ventral aspect; 521: aedeagus from the left; 522: aedeagus from the right; 523: caudal part of female abdomen from below. Scale: for 523 0.25 mm, for the rest 0.1 mm.

167

Key to species of *Xanthodelphax*

1   Appendages of male anal tube long, several times as long as broad. In Poland and German D.R.   *xanthus* Vilbaste, 1965
   – Appendages of anal tube in male short, triangular, about as long as broad   2
2 (1) Genital styles long, approximately half as long as height of pygofer (Text-fig. 524). Appendages of male anal tube more closely approximated. Lateral lobes of female basally smoothly narrowing (Text-fig. 532). Genital scale of female small but distinct, black, visible as a small dark point in front of ovipositor (Text-fig. 532)
63. *flaveolus* (Flor)
   – Styles short, length approximately 1/3 of height of pygofer (Text-fig. 534). Appendages of male anal tube widely apart. Lateral lobes of female basally abruptly narrowing (Text-fig. 541). Genital scale of female not pigmented, imperceptible
64. *stramineus* (Stål)

63. *Xanthodelphax flaveolus* (Flor, 1861)
   Plate-fig. 31, text-figs. 524–533.

Delphax flaveola Flor, 1861: 72.

Brachypters entirely light yellow without dark markings, only claws, spines on hind tibiae and on tarsi black, appendages of anal tube and apices of genital styles in male fuscous, genital scale of female (Text-fig. 533) black or fuscous. Abdomen in macropters sometimes with dark patches. Fore wings of brachypters hyaline, apically truncate or rounded, covering only abdominal terga I–IV. Fore wings of macropters transparent or whitish, semi-transparent, length less than twice the length of abdomen, veins towards apices more or less distinctly fuscous. Male pygofer as in Text-figs. 524, 525, male anal tube as in Text-figs. 526, 527, genital styles as in Text-figs. 528, 529, aedeagus as in Text-figs. 530, 531. Venter of caudal part of female abdomen as in Text-fig. 532. Length of brachypters 2.0–3.0 mm, of macropters (with wings) 3.0–3.5 mm.

Distribution. Denmark: NEZ, Dyrehaven in June (leg.?). LFM, Sundby Storskov in July (O. Jacobsen). – Widespread and locally common in Sweden, found in Sk., Bl., Sm., Gtl., Ög., Upl., Vstm. Ång., Vb., P. Lpm. – Norway: Os, Bø, TEy, TEi, SFi. – Rare and sporadic in East Fennoscandia, found in Ab, N, Ta, Oa, Sb, Kb, ObN, and Kr. – Widespread in Europe, also recorded from w. Siberia.

Biology. In meadows (Kuntze, 1937). "Auf einer trockenen Hangwiese mit reicher Vegetation" (Kontkanen, 1938). In both xerophilous and mesophilous meadow (Schiemenz, 1969). "In dry and mesic grass leys, in pastures, cereal fields, and surrounding wasteland bearing meadow vegetation" (Raatikainen & Vasarainen, 1976). Schiemenz (1969) places *X. flaveolus* provisionally among "Ei-Überwinterer", but Strübing (1956) states that "alle Arten der Gattung *Calligypona* sensu Ossiannilsson im dritten bis vierten Larvenstadien überwintern". This would include the present species. In Sweden, adults have been collected in June, July, and August.

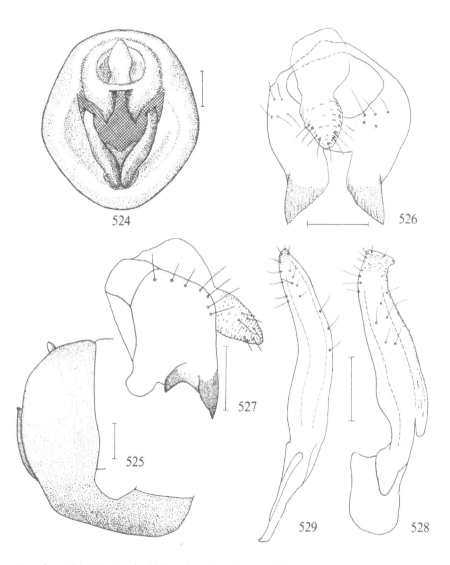

Text-figs. 524–529. *Xanthodelphax flaveolus* (Flor). – 524: male pygofer from behind; 525: male pygofer from the right; 526: male anal tube from behind; 527: male anal tube from the left; 528: left genital style from behind; 529: left genital style in lateral aspect. Scale: 0.1 mm.

64. Xanthodelphax stramineus (Stål, 1858)
Text-figs. 534–542.

Delphax straminea Stål, 1858: 358.

Very like the preceding species, differing only by the structure of male and female genitalia. The quotient length: width of fore wings in brachypters is on an average a little higher in *stramineus* than in *flaveolus*, but this is not a reliable character, as there is considerable overlapping in the ranges of variation of both species. Pygofer of male as in Text-figs. 534, 535, male anal tube as in Text-figs. 536, 537, genital style as in Text-fig.

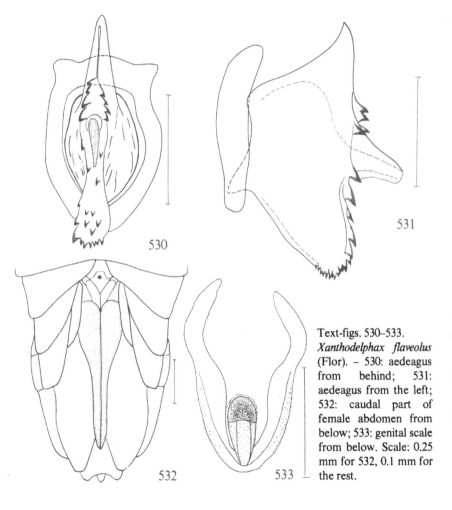

Text-figs. 530–533. *Xanthodelphax flaveolus* (Flor). – 530: aedeagus from behind; 531: aedeagus from the left; 532: caudal part of female abdomen from below; 533: genital scale from below. Scale: 0.25 mm for 532, 0.1 mm for the rest.

538, aedeagus as in Text-figs. 539, 540. Ventral aspect of posterior part of female abdomen as in Text-fig. 541, genital scale as in Text-fig. 542. Overall length of brachypters 2.0–3.0 mm, of macropters 3.35–4.0 mm.

Distribution. Rare in Denmark, found in LFM, Guldborgsund 12.VIII.1918, and in Bøtø 7.VII.1915 by O. Jacobsen, and in NEJ, Thorup strand 29.VI.1973 by E. Bøggild. – In Sweden less common than *flaveolus*, established in Sk., Sm., Öl., Ög., Vg., Boh.,

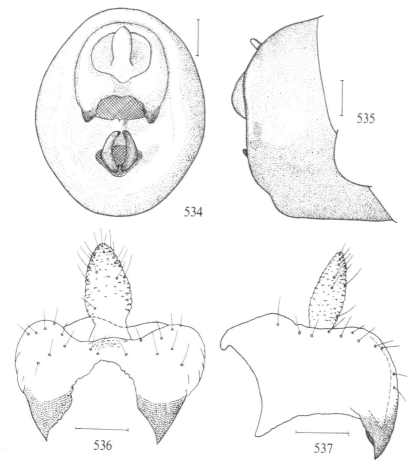

Text-figs. 534–537. *Xanthodelphax stramineus* (Stål). – 534: male pygofer from behind; 535: male pygofer from the right; 536: male anal tube from behind; 537: male anal tube from the left. Scale: 0.1 mm.

171

Dlsl., Sdm., Upl., and Äng. – So far not found in Norway. – Common in East Fennoscandia: Al, Ab, N – Om and Ok, also in Vib and Kr. – Widespread in Europe, also found in Kazakhstan.

Biology. On meadows (Kuntze, 1937). In dryish fields, moist sloping meadows, and peaty meadows (Linnavuori, 1952). "Mainly found in similar but perhaps slightly drier habitats than the preceding [*X. flaveolus*] species" (Raatikainen & Vasarainen, 1976). Hibernation takes place in the larval stages (Strübing in Müller, 1957; Remane, 1958). Sometimes found in mixed populations with the preceding species. In Sweden and Finland adults have been collected in June, July, and August.

Genus *Paradelphacodes* W. Wagner, 1963

Paradelphacodes W. Wagner, 1963: 169.

Type-species: *Delphax paludosa* Flor, 1861, by original designation.

Vertex as long as broad. Frons about 2 1/2 times as long as broad, broadest near middle, sides faintly convex or nearly parallel. Median carina of frons sharp but tending to be

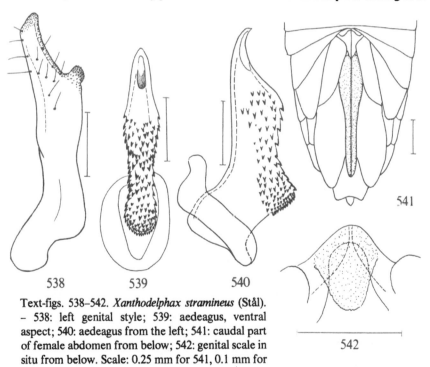

Text-figs. 538–542. *Xanthodelphax stramineus* (Stål). – 538: left genital style; 539: aedeagus, ventral aspect; 540: aedeagus from the left; 541: caudal part of female abdomen from below; 542: genital scale in situ from below. Scale: 0.25 mm for 541, 0.1 mm for the rest.

obsolescent on junction with vertex. Rostrum reaching to basis of hind coxae. Lateral carinae of pronotum curving outwards, not reaching hind border. Fore wings of brachypters apically rounded. Post-tibial calcar comparatively long, with many (>20) marginal teeth, apical tooth present. Male genital styles broadening towards apex. Appendages of anal tube in male short, parallel. In Europe one species.

65. *Paradelphacodes paludosa* (Flor, 1861)
Text-figs. 543–550.

Delphax paludosa Flor, 1861: 34.

Brownish yellow, abdomen of male blackish. Frons sometimes with a narrow light transverse band on lower margin. Antennae comparatively long, apex of first and basis of second segment dark. Fore wings of brachypters as long as abdomen, 2.1–2.5 times as long as broad, brownish yellow with darker, distinctly granulate veins. Fore wings of macropters twice as long, also with dark veins. Male pygofer as in Text-figs. 543, 544, anal tube of male as in Text-figs. 545, 546, genital styles as in Text-fig. 547, aedeagus as in Text-figs. 548, 549. Venter of posterior part of female abdomen as in Text-fig. 550. Overall length of brachypters 1.9–3.0 mm, of macropters 3.25–4.0 mm.

Distribution. Rare in Denmark: NWJ, Isbjerg, Hansted reservation 30.VII.1962 (N. P. Kristensen); B, Bastemose 21.VI.1976 (L. Trolle). – Rare also in Sweden, found in Sk., Sm., Ög., Upl., and Äng. – Not found in Norway. – Rare and sporadic in East Fennoscandia, established in Ab, Ta, Oa, Sb, Kb, and Kr. – Widespread in Europe (not in the Pyrenean Peninsula), also found in Mongolia, Maritime Territory, and Japan.

Biology. In wet marshes with *Carex*. "In places where sedges are growing. It lives very near the bog surface, often among *Sphagnum . . .*" (Linnavuori, 1952). In the "Cariceto canescentis – Agrostidetum caninae-association" (Marchand, 1953). On *Molinia* (Schiemenz, 1971). "In swamps, particularly tall-sedge bogs, and also in shore meadows" (Raatikainen & Vasarainen, 1976). Hibernation takes place in the larval stage (Remane, 1958). In Sweden, adults have been found in June and July.

Genus *Oncodelphax* W. Wagner, 1963
Oncodelphax W. Wagner, 1963: 169.
Type-species: *Delphax pullula* Boheman, 1852, by original designation.

Vertex as long as broad. Frons about 1.7 times as long as broad, broadest near middle, sides convex, median carina sharp but obsolescent on junction with vertex. Antennae short, first segment about as long as broad. Rostrum reaching basis of hind coxae. Lateral carinae of pronotum curved outwards, not reaching hind margin. Fore wings of brachypters apically rounded or truncate with rounded corners, 1.3–1.5 times as long as broad. Post-tibial calcar comparatively long, with 12–17 marginal teeth, apical tooth

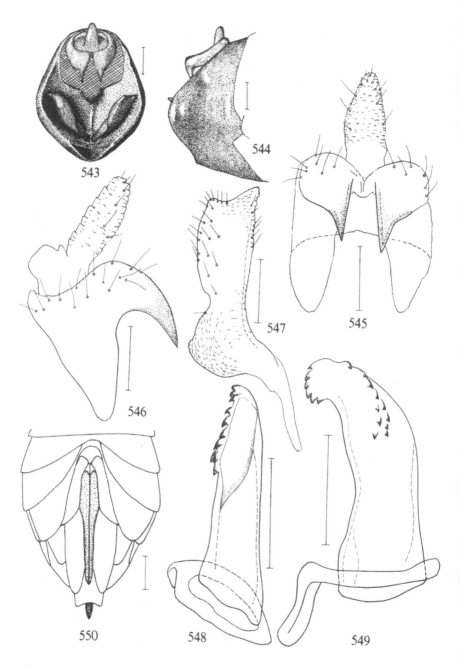

543

544

545

546

547

550

548

549

small. Male pygofer as seen from behind almost twice as high as broad. Appendages of male anal tube large, more than twice as long as basal width. Lateral lobes of female caudally protruding towards middle line. A distinct genital scale absent. One palaearctic species.

66. Oncodelphax pullulus (Boheman, 1852)
Plate-fig. 19, text-figs. 551–558.

Delphax pullula Boheman, 1852b: 116.

Head in males brownish yellow to light brown, clypeus a little lighter. Pronotum and mesonotum of brachypterous male light yellow, thoracal venter blackish brown, fore wings shining brown to black, apical margin narrowly yellowish, femora and tibiae yellowish, apices of tarsi fuscous. Abdomen yellowish, pygofer black. Brachypterous female dirty yellow or brownish yellow, fore wings concolorous, head often chestnut brown, clypeus lighter. Macropterous female often dark brown. Male pygofer as in Text-figs. 551, 552, male anal tube in lateral aspect as in Text-fig. 553, styles short (Text-fig. 554), aedeagus slender, almost straight (Text-figs. 555–557). Venter of caudal part of female abdomen as in Text-fig. 558. Length of brachypterous male 1.6–1.9 mm, of brachypterous female 2.1–2.4 mm, of macropterous female (with wings) 3.1–4.0 mm. I have not seen macropterous males of this species.

Distribution. Rare in Denmark: NWJ, Nors sø 25.VII.1962 (N. P. Kristensen); NEZ, Hillerød 6.VIII.1918 (O. Jacobsen); LFM, Marielyst forest 26.VII.1915 (O. Jacobsen). – Not rare, locally abundant in Sweden, established in Sk., Bl., Sm., Öl., Gtl., Ög., Vg., Upl., Vrm., Dlr. – Rare in Norway, found in AK, Oslo, Ormøya, according to Siebke (1874), and in HEs: Eidskog 11.VII.1974 (Ossiannilsson, 1977). – Scarce and sporadic in East Fennoscandia, found in Al, Ab, N, St, Ta, Oa, Kb, Om; Vib, Kr. – Austria, France, German D.R. and F.R., England, Scotland, Netherlands, Poland, Switzerland, Estonia, n. Russia, Yugoslavia.

Biology. In marshes with sedges (Kuntze, 1937). In "Carex vesicaria-Uferweissmohr" (Kontkanen, 1938). Tyrphobiont, in tall-sedge bogs and wet "rimpi" bogs (Linnavuori, 1952). In Sweden adults were collected in ult. May and in June and July.

Text-figs. 543–550. *Paradelphacodes paludosa* (Flor). – 543: male pygofer from behind; 544: male pygofer from the right; 545: male anal tube from behind; 546: male anal tube from the left; 547: right genital style; 548: aedeagus, ventral aspect; 549: aedeagus from the left; 550: caudal part of female abdomen from below. Scale: 0.25 mm for 550, 0.1 mm for the rest.

175

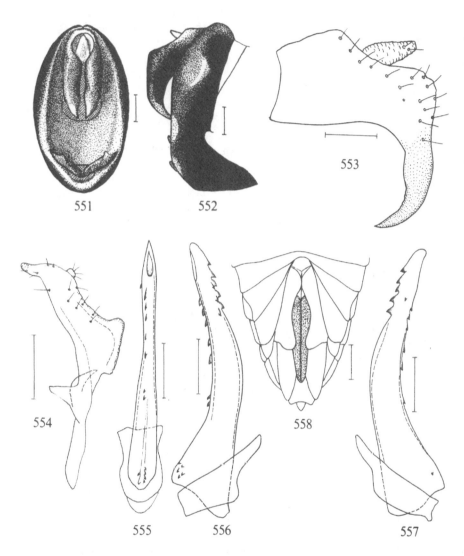

Text-figs. 551–558. *Oncodelphax pullulus* (Boheman). – 551: male pygofer from behind; 552: male pygofer from the right; 553: male anal tube from the left; 554: left genital style; 555: aedeagus, ventral aspect; 556: aedeagus from the right; 557: aedeagus from the left; 558: caudal part of female abdomen from below. Scale: 0.25 mm for 558, 0.1 mm for the rest.

Genus *Criomorphus* Curtis, 1833

Criomorphus Curtis, 1833: 195.
Type-species: *Criomorphus albomarginatus* Curtis, 1833, by original designation.

Frons with two median carinae. Vertex as broad as long or broader than long, sides approximately parallel. Antennae short. Side carinae of pronotum curved outwards, not reaching hind border. Fore wings of brachypters apically truncate, about 1.25 times as long as broad, shining black or brownish, hind margin broadly whitish. Marginal teeth of post-tibial calcar small, number small and much varying, apical tooth inconspicuous or absent. Pygofer of male without a lateral incision. Genital styles diverging, pointed. Genital scale of female distinct, pigmented. In Northern Europe three species.

Key to species of *Criomorphus*

1   Median carinae of frons usually separate throughout. Pygofer of male (Text-fig. 559) as seen from behind almost triangular, lower corners angular. Appendages of anal tube in male very short. Median margins of laterally lobes in female distally of middle each with a small tooth (Text-fig. 566). Genital scale of female (Text-fig. 567) small, width less than 0.2 mm   67. *albomarginatus* Curtis
-   Median carinae of frons confluent towards clypeus, forming a V or an Y. Pygofer of male not as above. Appendages of anal tube longer. Genital scale of female larger, width over 0.25 mm   2
2 (1) Carinae of frons distinct throughout. Appendages of anal tube in male thin (Text-fig. 576). Median margin of lateral lobe in female without a tooth (Text-fig. 283). Width of genital scale in female about 0.27 mm (Text-fig. 584)
                                                                        69. *borealis* (J. Sahlberg)
-   Carinae of frons evanescent towards junction with vertex. Appendages of anal tube in male stout (Text-fig. 568). Median margin of lateral lobe in female with a small tooth near middle (Text-fig. 574). Genital scale of female large, width about 0.43 mm. (Text-fig. 575)   68. *moestus* (Boheman)

67. *Criomorphus albomarginatus* Curtis, 1833
Plate-fig. 20, text-figs. 559–567.

Criomorphus albomarginatus Curtis, 1833: 195.
Delphax collaris Stål, 1853: 175.
Delphax adelpha Flor, 1861: 81.

Head and notum brownish yellow. Carinae of frons ivory-white, margined with black. Caudal border of pronotum broadly whitish. Posterior side-margins of pronotum also whitish ($\male$) or light yellowish ($\female$). Fore wings of brachypters reaching a little beyond hind border of third abdominal tergum, brownish black ($\male$) or brownish yellow ($\female$),

177

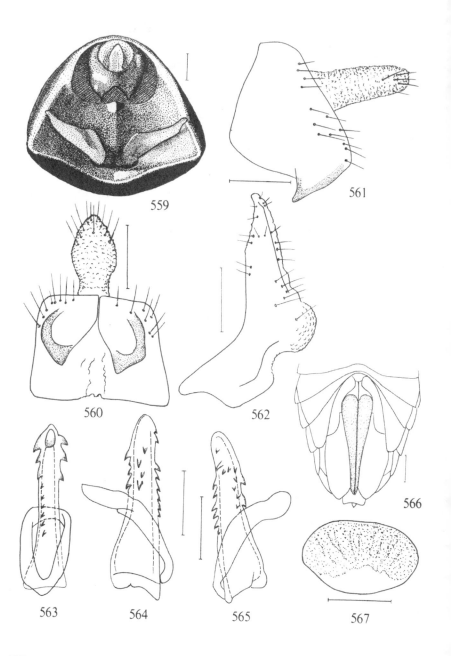

559

561

560

562

566

563 564 565 567

apical margin in both sexes broadly whitish. Wings in macropters transparent, colourless or sordid yellow, veins yellow. Venter in male largely blackish brown, in female brownish yellow. Abdomen in male black, in female brownish yellow, fore borders of tergal segments blackish, hind borders lighter or concolorous. Legs with dark longitudinal streaks. Macropters sometimes largely brownish yellow, sometimes more or less resembling brachypters in colour. Male pygofer as in Text-fig. 559, appendages of male anal tube short (Text-figs. 560, 561), genital style as in Text-fig. 562, aedeagus as in Text-figs. 563–565. Venter of posterior part of female abdomen as in Text-fig. 566, genital scale as Text-fig. 567. Overall length of brachypters 2–3.1 mm, of macropters 3.5–4 mm.

Distribution. Fairly common in Denmark, found in all districts except SJ. – Fairly common in Sweden, Sk. – Ång. – In Norway only found in Ø: Frederikstad and Hvaler (Helliesen); Bø: Bingen, Modum (Holgersen), and TEi: Øverland, Vestfjorddalen (Holgersen). – Fairly common but sporadic in South and Central East Fennoscandia, up to Oa, Sb, and Kb; also in Kr. – Widespread in Europe, also in Tunisia and Israel.

Biology. In meadows and coastal dunes and on low vegetation in forests (Kuntze, 1937). "Scattered in leys and in the undergrowth of the surrounding woods, from where it migrated to cereals" (Raatikainen & Vasarainen, 1973). "Nowadays, chiefly in leys, pastures, meadows at forest edges and cereal fields. In cages this species reproduced on oats and on *Festuca pratensis* and fed on *Poa pratensis, Phleum pratense, Agropyron repens* and *Deschampsia caespitosa*. (Raatikainen & Vasarainen, 1976). Hibernation takes place in the larval stage (Müller, 1957). In Sweden adults have been collected in May, June, and July. Macropters are rare.

68. *Criomorphus moestus* (Boheman, 1847)
   Text-figs. 568–575.

Delphax moesta Boheman, 1847a: 59.
Delphax thoracica Stål, 1858: 356.

Resembling the preceding species, differing by characters given in the key and by other details in the male genitalia. Carinae of frons usually not distinctly lighter than interspaces. Pygofer of male (Text-fig. 568) black, hind border fairly broadly whitish especially in upper part. Appendices of male anal tube (Text-fig. 569) thick and comparatively long. Genital style as in Text-fig. 570. Aedeagus (Text-figs. 571–573) slightly S-curved. Venter of posterior part of female abdomen as in Text-fig. 574, genital scale

---

Text-figs. 559–567. *Criomorphus albomarginatus* Curtis. – 559: male pygofer from behind; 560: male anal tube from behind; 561: male anal tube from the left; 562: left genital style; 563: aedeagus, ventral aspect; 564: aedeagus from the left; 565: aedeagus from the right; 566: caudal part of female abdomen from below; 567: genital scale from below. Scale: 0.25 mm for 566, 0.1 mm for the rest.

179

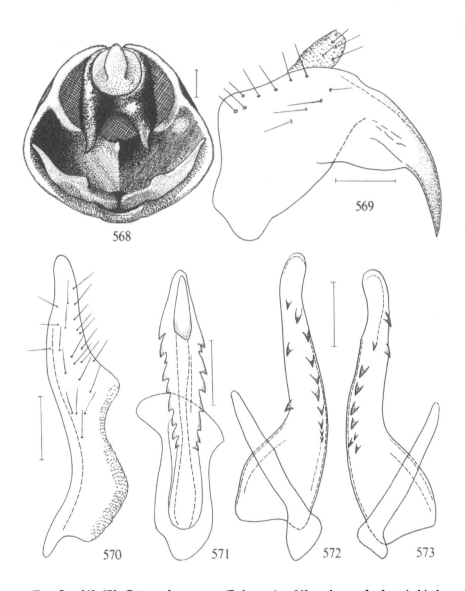

Text-figs. 568–573. *Criomorphus moestus* (Boheman). – 568: male pygofer from behind; 569: male anal tube from the left; 570: left genital style from behind; 571: aedeagus, ventral aspect; 572: aedeagus from the left; 573: aedeagus from the right. Scale: 0.1 mm.

as in Text-fig. 575. Overall length of brachypters 2.25–3.6 mm, of macropters 3.8–4.2 mm.

Distribution. Not found in Denmark, nor in Norway. – Probably not uncommon in Sweden in its normal biotope, so far established in Ög., Upl., Dlr., Hls., Ång., Nb., and Lu. Lpm. – Rare and sporadic in East Fennoscandia, found in N, Oa, Om, ObN, and LkW, also in Vib and Kr. – England, France, German D.R., Estonia, Latvia, Bohemia, n. Russia.

Biology. "Bewohnt die offenen seggenreichen Uferweissmoore" (Kontkanen, 1949). On the lower parts of *Calamagrostis canescens* (Ossiannilsson, 1944). Hibernation takes place in the larval stage (Müller, 1957). In Sweden, adults have been found in June and July. Macropters are rare.

69. *Criomorphus borealis* (J. Sahlberg, 1871)
Text-figs. 576–584.

Ditropis borealis J. Sahlberg, 1871: 477.

Resembling *Criomorphus albomarginatus*, differing by characters mentioned in the key, and by other details in the structure of the male genitalia. Pygofer of male (Text-fig. 576) black with yellowish margins. Anal tube of male as in Text-figs. 577, 578, genital style as in Text-fig. 579, aedeagus (Text-figs. 580–582) slightly curved. Ventral aspect of caudal part of female abdomen as in Text-fig. 583, genital scale as in Text-fig. 584. Length of brachypters 2.25–2.8 mm. I have not seen macropters of this species.

Distribution. Not found in Denmark. – In Sweden only in the north (Dlr. – T. Lpm.). – Norway: only found in Os, Ringebu (H. Holgersen), and Nsi: Saltdalen (J. Sahlberg). East Fennoscandia: common in the northern part, fairly common also in the south. – Estonia, Latvia, Poland, German D.R., Bohemia, n. Russia, n.w. Siberia, Mongolia.

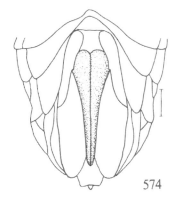
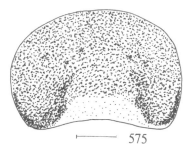

574

575

Text-figs. 574, 575. *Criomorphus moestus* (Boheman). – 574: caudal part of female abdomen from below; 575: genital scale from below. Scale: 0.25 mm for 574, 0.1 mm for 575.

181

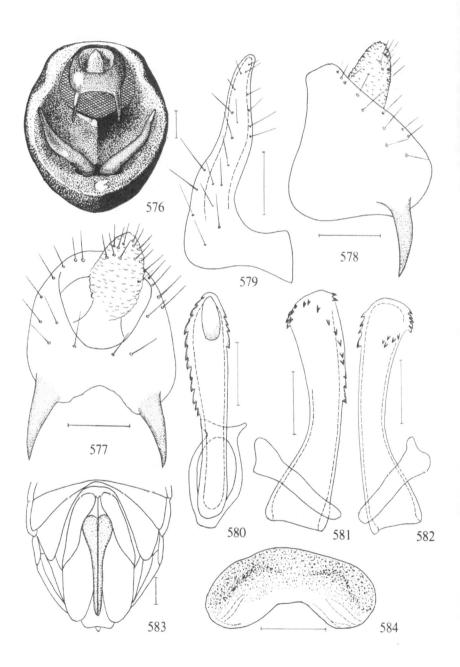

576

579

578

577

580

581

582

583

584

182

Biology. "Auf krautreichen Wiesen, hauptsächlich in der subalpinen Region und in krautreichen Birkenwäldern" (Lindberg, 1932a). "Bewohner der Feldschicht verschiedener mehr oder minder frischer Wälder und grasreicher Brücher" (Kontkanen, 1949). "Auf feuchten Waldwiesen ... im *Calamagrostis canescens* – *Molinia*-Bestand" (Schiemenz, 1976). Strübing (1960) reared *Criomorphus borealis* on *Calamagrostis canescens*. In Petsamo adults were found in June and July (Lindberg, 1932a). Our Swedish material comprises adults collected in June, July and August, also in Ab: Sammatti the species was found as late as 16.VIII (Lindberg, 1947). On the other hand, Strübing (1960) found "ältere Imagines" already on 30.V. (1959) in Grünewald, German D.R. Macropterous individuals are very rare.

Genus *Javesella* Fennah, 1963

Javesella Fennah, 1963: 15.
Type-species: *Fulgora pellucida* Fabricius, 1794, by original designation.
Weidnerianella W. Wagner, 1963: 170.
Type-species: *Fulgora pellucida* Fabricius, 1794, by original designation.
"Coarsely built. Vertex quadrate, parallel sided, anterior third surpassing eyes; frons about twice as long as broad; lateral pronotal carinae not reaching hind margin; calcar many-toothed. Pygofer rather elongate, medioventrally shallowly indented, lateral margins entire; diaphragm dorsally V-shaped, unarmed medially. Genital styles simple, sinuate, tapering, strongly divergent... First valvifers obtusely excavate near base" (Fennah, l. c.). Appendages of anal tube in male in lateral aspect hook-shaped, set close to each other, almost contiguous at base. In Denmark and Fennoscandia 9 species. Separation of females may be difficult.

Key to subgenera of *Javesella*

1   Appendages of anal tube in male short, usually strongly curved. Genital styles of male proximally of apex more or less compressed. Aedeagus either not curved towards venter, or hook-like (Text-fig. 592), not evenly arched
                                                                subgenus *Javesella* Fennah
–   Appendages of anal tube in male long, fairly curved. Genital styles not compressed near apex. Aedeagus evenly curved towards venter (Text-fig. 668)
                                                                subgenus *Haffnerianella* W. Wagner.

Text-figs. 576–584. *Criomorphus borealis* (J. Sahlberg). – 576: male pygofer from behind; 577: male anal tube from behind; 578: male anal tube from the left; 579: right genital style; 580: aedeagus, ventral aspect; 581: aedeagus from the left; 582: aedeagus from the right; 583: caudal part of female abdomen from below; 584: genital scale from below. Scale: 0.25 mm for 583, 0.1 mm for the rest.

Key to species of *Javesella*

1	Males	2
–	Females	11

2 (1) Inner margin of dorsal incision of pygofer caudally converging, forming a forceps-shaped figure (Text-fig. 648)   3
- Inner margin of dorsal incision of pygofer not converging caudally   4

3 (2) General colour of head and thorax dark, black or dark brownish, fore wings of brachypters blackish brown, basally lighter. Aedeagus as in Text-figs. 651–653
    77. *forcipata* (Boheman)
- General colour of head and thorax light, fore wings of brachypters light, sordid yellow or light yellow. Aedeagus as in Text-figs. 657–659 78. *alpina* (J. Sahlberg)

4 (2) Hind margin of pygofer in lateral aspect strongly convex (Text-fig. 625)   5
- Hind margin of pygofer in lateral aspect truncate   7

5 (4) Lateral outline of aedeagus as in Text-figs. 629, 630   74. *discolor* (Boheman)
- Lateral outline of aedeagus different   6

6 (5) Lateral outline of aedeagus as in Text-fig. 636   75. *simillima* (Linnavuori)
- Lateral outline of aedeagus as in Text-figs. 643, 644   76. *bottnica* Huldén

7 (4) Aedeagus recurved, in lateral aspect hook-like (Text-figs. 591, 592)
    70. *pellucida* (Fabricius)
- Aedeagus straight or evenly curved, not hook-like   8

8 (7) Aedeagus in lateral aspect forked into two lobes (Text-figs. 602, 603, 612, 620)   9
- Aedeagus not forked   79. *stali* (Metcalf)

9 (8) Ventral lobe of aedeagus approximately half as long as dorsal lobe (Text-fig. 620)   73. *salina* (Haupt)
- Ventral lobe of aedeagus as long as dorsal lobe or a little shorter   10

10 (9) Aedeagus deeply forked, lobes several times as long as broad (Text-figs. 602, 603)   71. *dubia* (Kirschbaum)
- Aedeagus shallowly forked, lobes 1 1/2–2 times as long as broad (Text-fig. 612)
    72. *obscurella* (Boheman)

11 (1) Basal dilatation of lateral lobe almost angular (Text-figs. 593, 594). Fore wings of brachypters over twice as long as broad. Margins of genital scale (Text-figs. 595, 596) not distinctly thickened   70. *pellucida* (Fabricius)
- Basal dilatation of lateral lobe rounded or inconspicuous   12

12 (11) Fore wings of brachypters 1.2–1.7 times as long as broad   13
- Fore wings of brachypters 1.7–2.25 times as long as broad   17

13 (12) Fore wings of brachypters brown or fuscous   14
- Fore wings of brachypters colourless   16

14 (13) Fore wings of brachypters entirely brown, only apical margin lighter
    79. *stali* (Metcalf)
- Fore wings of brachypters basally lighter   15

15 (14) Median carina of frons obsolete on junction with vertex. Head and thorax largely black   77. *forcipata* (Boheman)

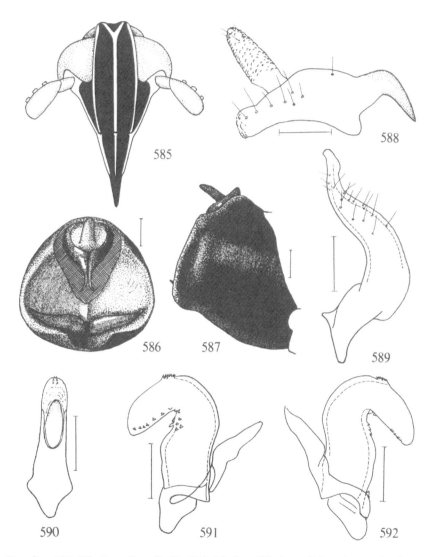

Text-figs. 585–592. *Javesella pellucida* (Fabricius). – 585: face; 586: male pygofer from behind; 587: male pygofer from the right; 588: male anal tube from the left; 589: left genital style; 590: aedeagus from behind; 591: aedeagus from the right; 592: aedeagus from the left. Scale: 0.1 mm.

185

–   Median carina of frons distinct throughout. Anterior part of body largely light
                                                                78. *alpina* (J. Sahlberg)
16 (13) Basal dilatation of lateral lobes large (Text-fig. 660). Genital scale small (Text-
    fig. 661)   78. *alpina* (J. Sahlberg)
–   Basal dilatation of lateral lobes less prominent
                                74. *discolor* (Boheman) & 75. *simillima* (Linnavuori)
17 (12) Fore wings of brachypters brownish, apical margin whitish. Genital scale as in
    Text-fig. 622   73. *salina* (Haupt)
–   Fore wings of brachypters colourless or smoke-coloured, apical margin con-
    colorous   18
18 (17) Median margin of lateral lobe faintly curved, basal dilatation inconspicuous
    (Text-fig. 645)   76. *bottnica* (Huldén)
–   Median margin of lateral lobe markedly curved, basal dilatation distinct (Text-
    fig. 605)   19
19 (18) Fore margin of genital scale strongly thickened (Text-fig. 606)
                                                                71. *dubia* (Kirschbaum)
–   Genital scale as in Text-fig. 614   72. *obscurella* (Boheman)

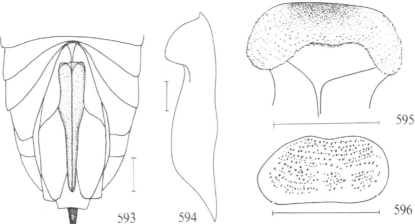

Text-figs. 593–596. *Javesella pellucida* (Fabricius). – 593: caudal part of female abdomen from below; 594: left lateral lobe of female from below; 595: genital scale in situ from below; 596: genital scale (another specimen) depressed under coverglass. Scale: 0.25 mm for 593, 0.1 mm for the rest.

Subgenus *Javesella* Fennah, 1963

70. *Javesella (Javesella) pellucida* (Fabricius, 1794)
    Plate-fig. 15, text-figs. 585–596.

Fulgora pellucida Fabricius, 1794: 7.
Fulgora marginata Fabricius, 1794: 7.
Delphax dispar Fallén, 1806: 126.
Liburnia flavipennis J. Sahlberg, 1871: 438.

Much varying in colour. Male (Plate-fig. 15) usually largely black, carinae of head (Text-fig. 585) yellowish, pronotum largely whitish. Tibiae dirty yellowish with black longitudinal streaks. Fore wings of brachypters 2.23–2.63 times as long as broad, apically rounded, colourless or yellowish, commissural margin often darkened. Fore wings of macropters colourless, transparent with partly darker veins. Especially brachypters often lighter in colour, partly brownish yellow. Female usually brownish yellow or sordid yellow, often with more or less extended blackish patches. Post-tibial calcar with 14–21 marginal teeth. Male pygofer as in Text-figs. 586–587, anal tube of male as in Text-fig. 588, genital style as in Text-fig. 589, aedeagus short, hook-like (Text-figs. 590–592). Ventral aspect of caudal part of female abdomen as in Text-fig. 593, left lateral lobe as in Text-fig. 594, genital scale as in Text-figs. 595, 596. (It must be remarked that the aspect of the genital scale as seen *in situ* often differs considerably from its appearance when it is depressed under a coverglass). Overall length of brachypters 2.1–3.4 mm, of macropters 4–5 mm.

Distribution. Common and found in almost all districts in Denmark, Sweden, and East Fennoscandia. – In Norway so far established in AK, HEn, On, Bø, Bv, VE, TEy, TEi, HOy, MRy, STi, NTy, Nsy, Nnø, TRi, and Fi. – Widespread in Europe, also in Algeria, Azores, Marocco, Anatolia, Altai, Kamchatka, Kazakhstan, Kirghizia, Kurile Islands, Libya, Maritime Territory, Mongolia, Sakhalin, Siberia, Uzbekistan.

Biology. On grasses in wet biotopes but also in cultivated fields. Owing to the economic importance of *J. pellucida,* its biology has been comparatively thoroughly studied, e. g. in Sweden and Finland (Tullgren, 1925; von Rosen, 1956; Kanervo & al., 1957; Heikinheimo, 1958; Jürisoo, 1964; Quayum, 1968). In Central Sweden and in the western coastal region of Finland the species is univoltine (von Rosen, 1956, Kanervo & al., 1957). Hibernation in Central Sweden takes place mainly in larval stages II and III (Tullgren, l. c., von Rosen, l. c.), in Finland mainly in stages III and IV (Kanervo & al., l. c.). Breeding plants and food-plants are many grasses, including cereals, *Avena sativa* and *Lolium perenne* being preferred as oviposition plants (Quayum, l. c.). Eggs are inserted into the hollow stems of the host-plants and can be found arranged in longitudinal rows inside these stems (Tullgren, l. c.). According to von Rosen (l. c.) one female can produce 500–1000 eggs. The average number of eggs laid in plants is 430 per female (Heikinheimo, l. c.), or 232 $\mp$ 62.28 (Quayum, l. c.). The incubation period lasts 2–3 weeks. Contrary to most other Delphacidae, macropters of *J. pellucida* are more common than brachypters in Denmark and Fennoscandia, as well as in England. But in Iceland brachypters dominate (Lindroth & al., 1973). According to results obtained by Quayum (l. c.) in laboratory studies, population density seems to have some effect on wing dimorphism relations, crowding in larval stages resulting in an increased proportion of macropterous females. But these conditions deserve more attention.

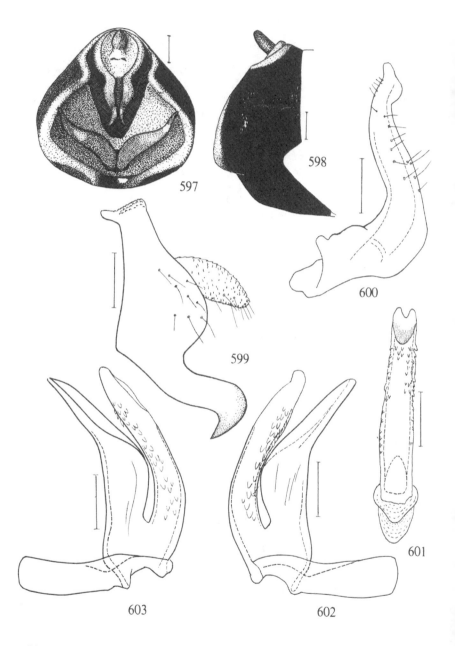

597

598

600

599

601

603

602

188

Economic importance. *Javesella pellucida* acts as a vector of two virus diseases of oats, viz. oat dwarf tillering disease (ODTD), and oat striate and red disease (OSRD) (Lindsten, 1961).

71. Javesella (Javesella) dubia (Kirschbaum, 1868)
Text-figs. 597–606.

Delphax dubia Kirchbaum, 1868: 26.
Delphax herrichii Kirschbaum, 1868: 26.
Delphax nitidipennis Kirschbaum, 1868: 31, partim.
Liburnia pargasensis Reuter, 1880: 197.
Liburnia difficilis Edwards, 1888: 197.

Resembling *pellucida* but brachypters are more common than macropters. Much varying in colour. The brachypterous male is often largely light but usually almost entirely black with carinae of head light; hind border of pronotum only narrowly light. Fore wings of brachypters 1.68–2.25 times as long as broad, colourless, yellowish or smoke-coloured. Macropterous male usually resembling the corresponding form of *pellucida*, pronotal hind border narrowly or faintly broadly whitish. The female varies accordingly. Male pygofer as in Text-figs. 597, 598, male anal tube as in Text-fig. 599, genital style as in Text-fig. 600, aedeagus (Text-figs. 601–603) longish, deeply forked. Venter of posterior part of female abdomen as in Text-fig. 604, left lateral lobe as in Text-fig. 605, anterior margin of genital scale (Text-fig. 606) arched, thickened. Overall length of brachypters 2.15–3.4 mm, of macropters 3.2–4.3 mm.

Distribution. Fairly common in Denmark, found in SJ, EJ, F, SZ, NWZ, NEZ, and B. – Common in Sweden, Sk. – P. Lpm. – Norway: VAy, Ry, Ri, HOi, MRy. – East Fennoscandia: found in most districts up to ObN and Ks, common at least in the southwest. – Widespread in Europe, also found in the Azores, Morocco, Altai, Kazakhstan, and Uzbekistan.

Biology. In marshes, forests and glades (Kuntze, 1937). At seashores, in moist meadows, sphagnous spruce woods, rich swampy woods, rich moist grass-herb woods, and moist *Oxalis-Myrtillus* spruce woods (Linnavuori, 1952). "Bewohner feuchter, schattiger Waldwiesen und Grasstellen" (Wagner & Franz, 1961). "In seashore and lakeshore meadows, swampy and moist heath forests, and especially open places in these, and in spruce-birch swamps. Nowadays chiefly in rather dry man-made meadows, pastures, leys and cereal fields. In tests it fed and reproduced on oats, and its natural host plants seem to be grasses" (Raatikainen & Vasarainen, 1976). Hibernation takes place in the larval stage (Remane, 1958). Adults in May–August.

Text-figs. 597–603. *Javesella dubia* (Kirschbaum). – 597: male pygofer from behind; 598: male pygofer from the right; 599: male anal tube from the left; 600: left genital style; 601: aedeagus, ventral aspect; 602: aedeagus from the right; 603: aedeagus from the left. Scale: 0.1 mm.

189

Economic importance. The species is a vector of European wheat striate mosaic virus (Kisimoto, 1961).

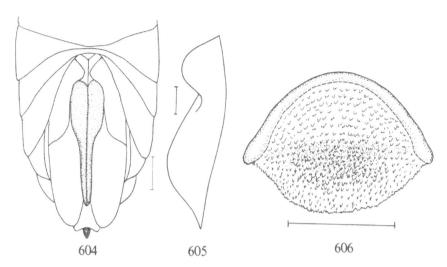

604   605   606

Text-figs. 604–606. *Javesella dubia* (Kirschbaum). – 604: caudal part of female abdomen from below; 605: left lateral lobe of female from below; 606: genital scale from below. Scale: 0.25 mm for 604, 0.1 mm for the rest.

72. *Javesella (Javesella) obscurella* (Boheman, 1847)
    Text-figs. 607–614.

Delphax obscurella Boheman, 1847a: 53.
Liburnia discreta Edwards, 1888: 197.

Resembling *pellucida* and *dubia*, brachypters more common than macropters. Colour much varying between brownish yellow and deep black, dark forms predominating. In these dark specimens the hind border of pronotum is only narrowly light and fore wings of brachypters may be blackish brown. Ratio length: width of fore wings in brachypters = 1.85–2.1. Veins of fore wings in both brachypters and macropters with conspicuous setigerous granules. In lateral aspect, hind border of male pygofer (Text-fig. 608) less straightly cut off than in *pellucida* and *dubia*, but less convex than in *discolor*. Male pygofer from behind as in Text-fig. 607, male anal tube as in Text-fig. 609, genital style as in Text-fig. 610. Aedeagus (Text-figs. 611, 612) short, forked but less deeply so than in *dubia* and *salina*. Venter of posterior part of female abdomen as in Text-fig. 613, genital scale as in Text-fig. 614. Overall length of brachypters 1.9–3.15 mm, of macropters 3.5–4.25 mm.

190

Distribution. Scarce in Denmark, found in EJ, NEJ, F, SZ, NEZ. – Common in Sweden, Sk. – T. Lpm. – The Norwegian records given by Siebke (1874) cannot be accepted without revision. H. Holgersen collected *obscurella* in a few localities in Ry, and Soot-Ryen found it in Nsy: Sømna, Sømnes. – East Fennoscandia: common, found up to LkW. – Holarctic. Widespread in Europe, also found in Anatolia, n. Siberia, and Mongolia.

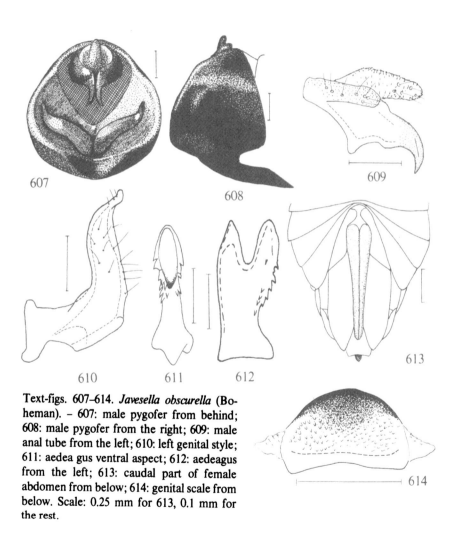

Text-figs. 607–614. *Javesella obscurella* (Boheman). – 607: male pygofer from behind; 608: male pygofer from the right; 609: male anal tube from the left; 610: left genital style; 611: aedea gus ventral aspect; 612: aedeagus from the left; 613: caudal part of female abdomen from below; 614: genital scale from below. Scale: 0.25 mm for 613, 0.1 mm for the rest.

191

Biology. In "Salzstellen, Flachmooren, Wäldern, Waldlichtungen, Wiesen" (Kuntze, 1937). Belongs to the "Molienetalia" (Marchand, 1953). Ikäheimo & Raatikainen (1961) gave a brief account of the life cycle of *J. obscurella* in Finland. According to these authors the species "produces one generation a year and hibernates at the nymphal stage. . . . Oviposition begins in June. The eggs are usually laid in the stems and leaves of cereals. . . . After the harvest, the nymphs feed on shoots remaining in the stubble, on gramineous weeds and on the timothy grass established in the cereals. The nymphs seem to spend the winter in the habitats where they hatched, on the surface of the ground. . . . Emergence starts in May or in some years not until June. – From the end of May to beginning of July the long-winged leafhoppers migrate from the grass leys, where they have overwintered, mainly to spring cereals and oviposit in these." In Sweden adults have been found in May, June, July, August, and September.

Economic importance. *Javesella obscurella* is a vector of the wheat striate mosaic virus and the oat sterile-dwarf virus (Ikäheimo & Raatikainen, 1961).

73. *Javesella (Javesella) salina* (Haupt, 1924)
Text-figs. 615–622.

Delphax salina Haupt, 1924: 298.
Liburnia juncea Haupt, 1935: 141.

Head sordid yellowish, frons between carinae concolorous or mottled with fuscous. Median carina of frons more or less obsolescent on junction with vertex. Pronotum largely sordid yellow or fuscous with hind border lighter. Mesonotum darker or lighter brown or fuscous, hind margin more or less distinctly lighter. Venter of thorax largely dark brown or sordid yellow with fuscous patches. Fore wings of brachypters apically rounded, usually dark brown, sometimes lighter, apical margin narrowly whitish. Ratio length: width of fore wings in brachypters = 1.7–2.0. Abdomen of male black. Legs usually largely sordid yellow. Female often lighter coloured. Pygofer of male as in Text-figs. 615, 616, male anal tube as in Text-fig. 617, genital style as in Text-fig. 618, aedeagus (Text-figs. 619, 620) deeply forked, lower branch about half as long as the upper one. Venter of posterior part of female abdomen as in Text-fig. 621, genital scale as in Text-fig. 622. Length of brachypters 1.88–2.6 mm, macropters about 3.4 mm.

Distribution. Very rare in Sweden, not found in Denmark, Norway, and East Fennoscandia. Håkan Lindberg found one male in Öl., Vickleby 18.IX.1946. Ossiannilsson collected 2 brachypterous specimens in Öl., Gårdslösa, Halltorp 12.VIII.1964 and seven brachypterous specimens at the seashore in Öl., Gårdslösa 13–15.VIII.1964. One macropterous male was found by the same collector in Gtl., Visby 8.VII.1952. – German D.R. and F.R., Poland, Estonia, Slovakia, Anatolia, Altai, Mongolia, Maritime Territory.

Biology. On *Juncus* (Haupt, 1935). On *Juncus Gerardi* in "Salzstellen" (Kuntze, 1937). According to Kuntze (1937) also on *"Brixa intermedia"* (= *Briza media?*) in

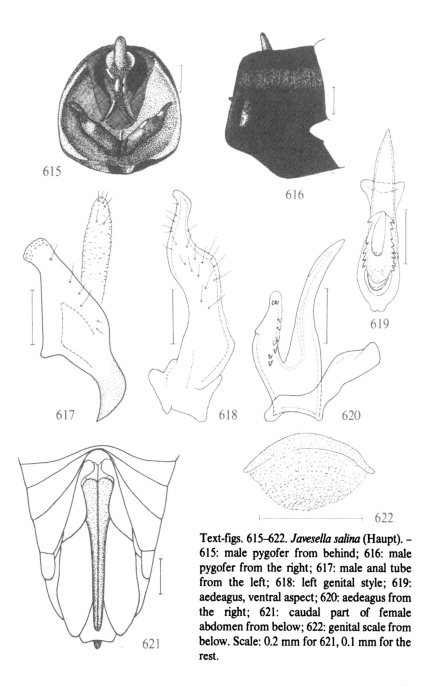

Text-figs. 615–622. *Javesella salina* (Haupt). –
615: male pygofer from behind; 616: male
pygofer from the right; 617: male anal tube
from the left; 618: left genital style; 619:
aedeagus, ventral aspect; 620: aedeagus from
the right; 621: caudal part of female
abdomen from below; 622: genital scale from
below. Scale: 0.2 mm for 621, 0.1 mm for the
rest.

"Flachmooren". As reported under "Distribution", the Swedish specimens were found in July, August, and September. Brachypters seem to be more common than macropters.

74. *Javesella (Javesella) discolor* (Boheman, 1847)
Text-figs. 623–633.

Delphax discolor Boheman, 1847a: 61.

Frons comparatively broader than in *pellucida* and *dubia* (Text-fig. 623; cf. Text-fig. 585). Body in both sexes usually largely black, carinae of head, hind border of pronotum and apex of scutellum, brownish yellow. Fore wings of brachypters apically rounded, sometimes almost colourless, sometimes brownish, in the latter case apical margin narrowly lighter. Index length: width of fore wings in brachypters 1.4–1.7. Females are often lighter, much varying in colour. Specimens from subarctic biotopes may also be more or less light-coloured. Male pygofer as in Text-figs. 624, 625, male anal tube as in Text-fig. 626, genital style as in Text-fig. 627, aedeagus as in Text-figs. 628–630, ventral aspect of caudal part of female abdomen as in Text-fig. 631, genital scale as in Text-figs. 632, 633. Overall length of brachypters 2.25–3.3 mm, macropters 4–4.5 mm.

Distribution. Common and widespread in Denmark, Sweden (Sk. – T. Lpm.) and East Fennoscandia (Al, Ab, N – Le and Li). – In Norway so far found in AK, HEn, Os, TEi, VAy, Ry, Ri, HOi, MRi, STi, Nsi, Nnø, TRy, and TRi. – Widespread in Europe, also found in Algeria, n. and m. Siberia, and Mongolia.

Biology. In glades (Kuntze, 1937). Tyrphobiont, in tall-sedge bogs and sphagnous spruce woods (Linnavuori, 1952). "In spruce-birch swamps, swampy meadows and moist forests, sporadically also in leys and cereal fields and near arable land" (Raatikainen & Vasarainen, 1962). Oats can serve as a host-plant (Vacke, 1962). In "bewaldetem soligenem Kesselmoor mit Randsümpfen. Vegetation: *Sphagnum recurvum, Eriophorum vaginatum*, Birke; *Cartex rostrata, Juncus effusus, Eriophorum angustifolium*" (Schiemenz, 1976). Univoltine. Hibernation takes place in the larval stage (Remane, 1958). In Sweden adults occur in May–September.

Economic importance. The species can transmit oat sterile-dwarf virus (Vacke, 1962), but its importance and ability as a vector of this disease has not been studied in Fennoscandia.

Text-figs. 623–633. *Javesella discolor* (Boheman). – 623: face; 624: male pygofer from behind; 625: male pygofer from the right; 626: male anal tube from the left; 627: left genital style from outside; 628: aedeagus, ventral aspect; 629: aedeagus from the left; 630: aedeagus from the right; 631: caudal part of female abdomen from below; 632: genital scale, depressed under coverglass; 633: genital scale (another specimen) in situ from below. Scale: 0.25 mm for 632, 0.1 mm for the rest.

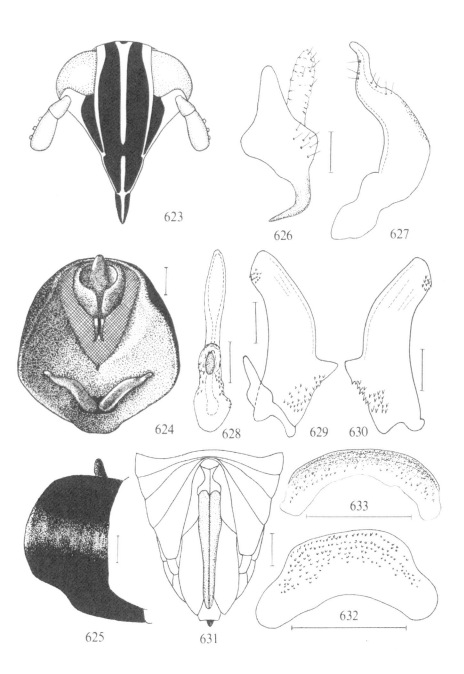

623

626 627

624 628 629 630

633

632

625 631

195

75. *Javesella (Javesella) simillima* (Linnavuori, 1948)
Text-figs. 634–639.

Liburnia nitidipennis J. Sahlberg, 1871: 452, p. p. (nec Kirschbaum, 1868).
Calligypona simillima Linnavuori, 1948: 45.

The following is a translation of Linnavuori's description in German. "♂: Shape as that of *C. discolor* (Boh.). Median carina and margins of frons and narrow margin of clypeus light yellow, interspaces dark brown. Vertex white, pronotum entirely and mesonotum

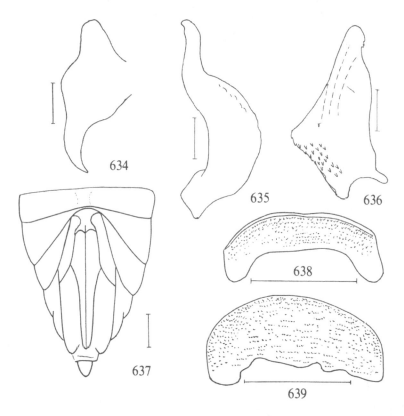

Text-figs. 634–639. *Javesella simillima* (Linnavuori). – 634: outline of anal tube (without anal style) from the right; 635: left genital style; 636: aedeagus from the right; 637: caudal part of female abdomen from below; 638: genital scale in situ from below (specimen from Taimyr); 639: genital scale depressed under coverglass (another specimen from Taimyr). Scale: 0.25 mm for 637, 0.1 mm for the rest. (637 after Vilbaste, 1971).

light yellow. Abdomen blackish with some yellow markings on segments 1 and 2. Fore wings uniform in width with rounded apices, about 1 1/2 times as long as broad, a little shorter than abdomen. Antennae and legs yellowish, pretarsus darker. Genital segment as in *discolor*, but genitalia distinctly different. ♀: As *C. discolor*, but body a little more slender. Median carina and margins of frons and clypeus yellow, interspaces on clypeus and on lower part of frons brownish, rest of body light yellow. Fore wings considerably shorter than abdomen, about 1 1/4 times as long as broad. Genitalia as in *C. discolor*. Length ♂ 1.5–1.6 mm; ♀ 2.0–2.1 mm. Macropterous form as yet ûnknown." Index length: width of fore wings in brachypterous males = 1.57–1.68 (Huldén, 1974). Lateral outline of male anal tube as in Text-fig. 634, genital style as in Text-fig. 635, aedeagus as in Text-fig. 636. Venter of posterior part of female abdomen as in Text-fig. 637, genital scale as in Text-figs. 638, 639.

Distribution. East Fennoscandia: Ab: Pargas (Reuter), Karislojo (J. Sahlberg), Raisio (Linnavuori); Ta: Ruovesi (J. Sahlberg); Sb: Kiuruvesi (Linnavuori); Kr: Jaakkima (J. Sahlberg). – Found in two localities in Estonia (Vilbaste, 1971). German D.R.: Erzgebirge; Thüringer Wald (Schiemenz, 1975); n. Siberia: Taimyr.

Biology. On *Eriophorum* (Vilbaste, 1971). On *Eriophorum* and *Carex* (Schiemenz, 1975).

76. *Javesella (Javesella) bottnica* Huldén, 1974
Text-figs. 640–645.

Javesella bottnica Huldén, 1974: 114.

The following is a transcript of the original description: ♂ length: 2.00–2.46 mm (f. brach.), 3.8 mm (f. macr.). Head black to blackish brown with pale yellowish brown carinae. Antennal segments yellowish brown with reddish brown bases. Pronotum with light yellow carinae and posterior edge, between the carinae varying in colour from black to light yellow, with dark patches. Scutellum blackish brown. Femora at base and distal segment of tarsi dark. Wings brownish. . . . ♀ length: 2.66 mm in f. brach., unknown in f. macr., similar to male, but with larger light patches and light spots on last three segments of abdomen. . . . Index of length and width of wing (short-winged males) 1.97 (1.79–2.32)." Male anal tube as in Text-fig. 640, genital style as in Text-fig. 641, aedeagus as in Text-figs. 642–644, ventral aspect of posterior part of female abdomen as in Text-fig. 645.

Distribution. So far only found in Finland and adjacent part of USSR. Ta: Ruovesi (J. Sahlberg); Kl: Parikkala (Hellén); Oa: Petalaks; Bergö; Maxmo (Håkan Lindberg); Sb: Jorois (P. H. Lindberg); ObN: Rovaniemi (Håkan Lindberg); LkW: Pallasjärvi (Wegelius); Kr: Dworetz (Günther), Paadana (J. Sahlberg).

Biology. Not studied. Adults were found in June and July. Macropters seem to be more rare than brachypters.

197

77. Javesella (Javesella) forcipata (Boheman, 1847)
Text-figs. 646–655.

Delphax forcipata Boheman, 1847a: 57.

Frons black between brownish yellow carinae, vertex dark brown, carinae obsolescent. Thorax in dark individuals sometimes entirely black without traces of lighter colour; usually the hind border is narrowly and indistinctly brownish yellow. Fore wings of brachypters black-brownish or dark brownish, proximally lighter. Index length: width of fore wings in brachypters = 1.20–1.63. Abdomen black. Male pygofer as in Text-figs. 646–648, anal tube of male as in Text-fig. 649, genital style as in Text-fig. 650, aedeagus as in Text-figs. 651–653, venter of caudal part of female abdomen as in Text-fig. 654, genital scale as in Text-fig. 655. Length of brachypters 2–3.5 mm, of macropters 3.3–4 mm.

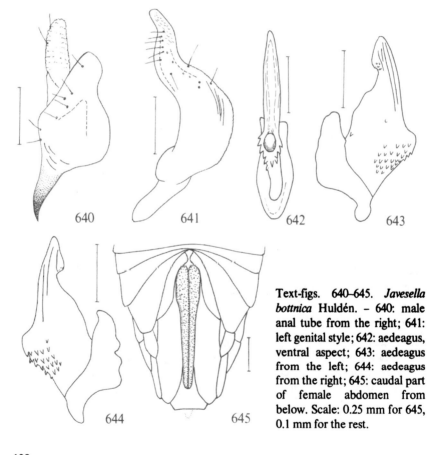

640 641 642 643

644 645

Text-figs. 640–645. *Javesella bottnica* Huldén. – 640: male anal tube from the right; 641: left genital style; 642: aedeagus, ventral aspect; 643: aedeagus from the left; 644: aedeagus from the right; 645: caudal part of female abdomen from below. Scale: 0.25 mm for 645, 0.1 mm for the rest.

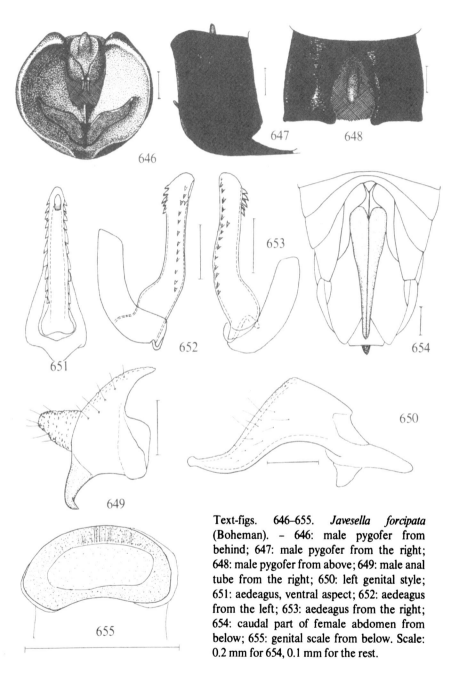

Text-figs. 646–655. *Javesella forcipata* (Boheman). – 646: male pygofer from behind; 647: male pygofer from the right; 648: male pygofer from above; 649: male anal tube from the right; 650: left genital style; 651: aedeagus, ventral aspect; 652: aedeagus from the left; 653: aedeagus from the right; 654: caudal part of female abdomen from below; 655: genital scale from below. Scale: 0.2 mm for 654, 0.1 mm for the rest.

Distribution. Fairly common in Denmark, found in SJ, EJ, F, LFM, NWZ, NEZ, B. - Common in Sweden, Sk. - T. Lpm. - Norway: VAy, HOi, SFi. - Common everywhere in East Fennoscandia. - Widespread in Europe, also recorded from n. Siberia (= *alpina*?).

Biology. In forests and meadows (Kuntze, 1937). "Bewohnt feuchte Wiesen und Moore" (Wagner & Franz, 1961). At watercourses with *Eriophorum* and *Carex* and in "*Eriophorum*-Hochmoor-Bultgesellschaften bzw. Zwischenmoor-Bultgesellschaften" (Schiemenz, 1975). "In swampy meadows, moist forests, grass-herb forests and spruce-birch swamps. Nowadays in field margins and on arable land, particularly in leys, and cereal fields" (Raatikainen & Vasarainen, 1976). Univoltine, hibernation in larval stage (Schiemenz, l. c., Raatikainen & Vasarainen, l. c.). We found adults in late May, June, July, and August. Brachypters are more common than macropters.

78. *Javesella (Javesella) alpina* (J. Sahlberg, 1871)
   Text-figs. 656–661.

Liburnia alpina J. Sahlberg, 1871: 462.

Closely related to *forcipata* but lighter in colour. Median carina of frons obtuse on junction with vertex, but fairly distinct throughout. Head and thorax, often also abdomen, largely yellow-brownish or dirty yellow with or without dark markings. Fore wings of brachypters yellow-brownish or dirty yellow, unicolorous or darker towards apex. Abdomen of male usually blackish. Ratio length: width of fore wings in brachypters = 1.48–1.7. Male genital style as in Text-fig. 656, aedeagus (Text-figs. 657–659) longer and comparatively more slender than in *forcipata*. Venter of posterior part of female abdomen as in Text-fig. 660, genital scale as in Text-fig. 661. Length of brachypters 2.15–3 mm. I have not seen macropters of this species.

Distribution. Not in Denmark. - Sweden: Hls., Nb., Ly. Lpm., Lu. Lpm., T. Lpm. - Norway: Nsy: Salten (J. Sahlberg); TRy: Skiervø (Håkan Lindberg). - Finland: Le: Vuontisjärvi (J. Sahlberg), Kilpisjärvi (Håkan Lindberg). - USSR: Lr: Kantalaks, Imandra; Petsamo (Håkan Lindberg); Kanin Peninsula (Poppius).

Biology. Among herbaceous vegetation (*Geranium, Vaccinium myrtillus, Trollius, Dryopteris, Phegopteris*) in pine and birch wood in regio silvatica and regio subarctica, with *Doliotettix lunulatus* and *Criomorphus borealis* (Lindberg, 1932a). In the north of Sweden adults were collected in June, July and August. Macropters apparently rare.

Subgenus *Haffnerianella* W. Wagner, 1966

Javesella Haffnerianella W. Wagner, 1966: 96.
Type-species: *Delphacodes stali* Metcalf, 1943, by original designation.

79. Javesella (Haffnerianella) stali (Metcalf, 1943)
Text-figs. 662–670.

Delphax bohemani Stål, 1858: 357 (nec 1854).
Delphacodes ståli Metcalf, 1943: 510 (nom. nov.).

Carinae of head distinct. Head and thorax sordid yellow or brownish yellow, more or less mottled with fuscous. Fore wings of brachypters 1.4–1.5 times as long as broad, brownish black, apical margin rounded, yellow-whitish. Fore wings of macropters 2/3

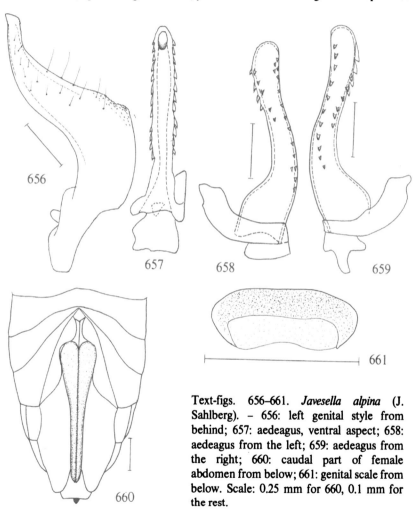

Text-figs. 656–661. *Javesella alpina* (J. Sahlberg). – 656: left genital style from behind; 657: aedeagus, ventral aspect; 658: aedeagus from the left; 659: aedeagus from the right; 660: caudal part of female abdomen from below; 661: genital scale from below. Scale: 0.25 mm for 660, 0.1 mm for the rest.

201

or 3/4 longer than abdomen, fuscous or sordid yellow with yellowish veins. Male abdomen black, dorsally with a median row of yellow-whitish spots, also laterally light spotted, hind border of pygofer yellow. Abdomen of female light yellow or dirty yellow, laterally more or less mottled with fuscous. Legs yellowish with blackish longitudinal streaks. Male pygofer as in Text-figs. 662, 663, male anal tube as in Text-fig. 664, genital style not constricted proximally of apex (Text-fig. 665), aedeagus as in Text-figs. 666–669. Venter of posterior part of female abdomen as in Text-fig. 670. Genital scale rudimentary, small, poorly sclerotized. Length of brachypters 2.3–3.5 mm, macropters 3.2–4.5 mm.

Distribution. So far not found in Denmark, nor in Norway. – Very rare in Sweden: Dlr: Krylbo 3.VII.1954 (Ossiannilsson); Ång.: Nyland, Gårdnäs and Hämra (Stål). – Scarce, but widespread in East Fennoscandia, found in Ab, N, Ta, Oa, Tb, Ks, LkE, Le, Li; Vib, Kr, Lr. – Austria, France, German D.R., ? Italy, Poland, Estonia, Latvia, n. Russia, Ukraine.

Biology. On *Equisetum* on sandy river-banks (Lindberg, 1932).

Genus *Ribautodelphax* W. Wagner, 1963

Ribautodelphax W. Wagner, 1963: 170.
Type-species: *Delphax collina* Boheman, 1847, by original designation.

Vertex approximately as long as broad. Carinae of head distinct throughout. Frons about twice as long as broad, broadest between lower part of eyes. Lateral carinae of pronotum only faintly curved, not reaching hind border. Vertex and notum with a more or less broad pale median line. Fore wings of brachypters apically rounded. Number of marginal teeth on post-tibial calcar between 10 and 20. Genital styles of male flattened, laterally widening in a flat lobe (reduced in *albostriatus*). Appendages of male anal tube spiniform, with a tendency of crossing each other. In Denmark and Fennoscandia five species, separation of females difficult.

Key to species of *Ribautodelphax*

1 Genital styles of male small, not widened in a flat lateral lobe (Text-fig. 706). Genital phragm without paired processes. Male pygofer in lateral aspect without an incision in upper part (Text-fig. 704). Body usually with strongly developed pigmentation, abdomen of brachypterous female with a dark pattern consisting of 3–4 pairs of longitudinal rows of spots. Genital scale of female as in Text-fig. 710
84. *albostriatus* (Fieber)
– Genital styles long, laterally widening in a flat lobe. Pygofer in lateral aspect with an incision in upper part. Genital phragm with a pair of small tooth-like processes. General pigmentation weaker, especially in females, abdomen of brachypterous females usually largely pale 2

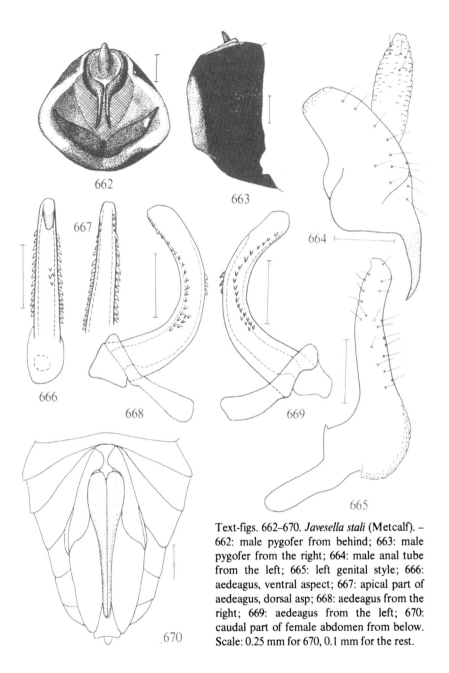

662

663

667

664

666

668

669

665

Text-figs. 662–670. *Javesella stali* (Metcalf). –
662: male pygofer from behind; 663: male
pygofer from the right; 664: male anal tube
from the left; 665: left genital style; 666:
aedeagus, ventral aspect; 667: apical part of
aedeagus, dorsal asp; 668: aedeagus from the
right; 669: aedeagus from the left; 670:
caudal part of female abdomen from below.
Scale: 0.25 mm for 670, 0.1 mm for the rest.

670

2 (1) Ratio length: width of fore wings in brachypters = 1.76–2.17. Ventral concave incision of male pygofer flanked by two pointed processes (Text-fig. 673). Genital scale of female as in Text-fig. 679   80. *collinus* (Boheman)
– Ratio length: width of fore wings in brachypters not over 1.9. Male pygofer ventrally without two distinct pointed projections   3
3 (2) Ventral concave incision in male pygofer weak, laterally indistinctly delimited
                    83. *pallens* (Stål)
– Ventral concave incision distinct, laterally delimited by angular corners   4
4 (3) Corners of ventral incision of pygofer very distinctly angular (Text-fig. 682). Appendages of male anal tube comparatively short, curved, neither of them vertical (Text-figs. 680, 683)   81. *angulosus* (Ribaut)
– Corners of ventral incision of pygofer less distinctly angular. Appendages of anal tube more elongate, one of them vertically suspended (Text-fig. 689)
                    82. *pungens* (Ribaut).

80. **Ribautodelphax collinus** (Boheman, 1847)
   Text-figs. 671–679.

Delphax collina Boheman, 1847a: 51.
Delphax concinna Fieber, 1866b: 525.
Liburnia biarmica J. Sahlberg, 1871: 430.

Head and thorax of brachypterous male straw-coloured with carinae of head whitish, black-edged, and a light median band on pro- and mesonotum. Fore wings pale straw-coloured, apically rounded. Legs straw-coloured with indistinct dark longitudinal streaks. Abdomen black with a light median stripe, often also laterally light-spotted, hind border of pygofer light. Brachypterous female largely pale yellow, with carinae of head black-edged, and with a white median band on pro- and mesonotum. The macropterous female is darker in colour. Fore wings of macropters in both sexes transparent with partly fuscous veins. Male pygofer as in Text-figs. 671, 672, ventral incision of male pygofer as in Text-fig. 673, anal tube of male as in Text-fig. 674, genital style as in Text-fig. 675, aedeagus as in Text-figs. 676, 677. Ventral aspect of caudal part of female abdomen as in Text-fig. 678, genital scale as in Text-fig. 679. Length of brachypters 2.4–3.3 mm, of macropters (with wings) 3.5–4 mm.

Distribution. Scarce in Denmark, found in SJ, EJ, NEZ, and B. – Not uncommon in southern and central Sweden, established in Sk., Bl., Hall., Sm., Öl., Ög., Upl., and Ång. – Norway: Holgersen collected *R. collinus* in Bø: Modum, Bingen, and in TEi: Seljord, and in TEi: Bø, Øvrebø. – Comparatively common in the south of East-Fennoscandia, scarce in the rest of Finland, found in Ab, N, Ta, Sa, Kb, Kr. – Widespread in Europe (not in Great Britain), also in Kazakhstan.

Biology. In woods and meadows (Kuntze, 1937). On "Binnendünen" (Remane, 1958). In Northern Germany on grasses among *Calluna* on sandy ground (Wagner & Franz, 1961). Stenotope species of dry grass meadows (Schiemenz, 1969). Hibernation

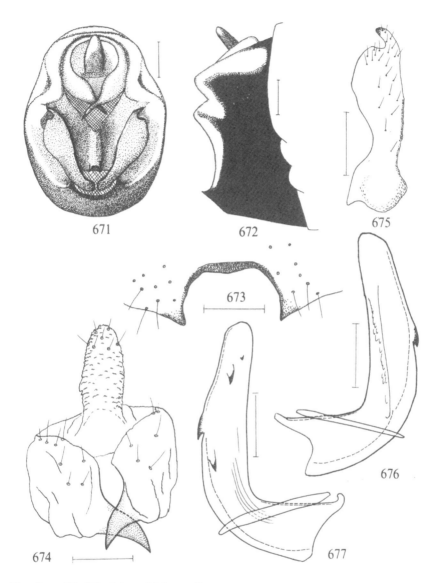

Text-figs. 671–677. *Ribautodelphax collinus* (Boheman). – 671: male pygofer from behind; 672: male pygofer from the right; 673: ventral margin of male pygofer from below; 674: male anal tube from behind; 675: left genital style; 676: aedeagus from the right; 677: aedeagus from the left. Scale: 0.1 mm.

in larval stage (Strübing, 1956, Müller, 1957, Remane, 1958, Schiemenz, 1969). We found adults in May–August. Brachypters are more common than macropters.

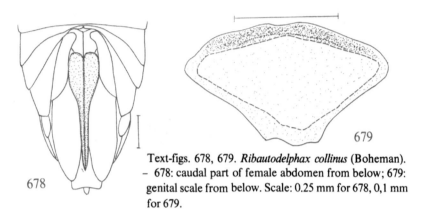

Text-figs. 678, 679. *Ribautodelphax collinus* (Boheman). – 678: caudal part of female abdomen from below; 679: genital scale from below. Scale: 0.25 mm for 678, 0,1 mm for 679.

81. **Ribautodelphax angulosus** (Ribaut, 1953)
    Text-figs. 680–688.

Calligypona angulosa Ribaut, 1953: 247.

Very much resembling *R. collinus*, differing by a slightly smaller body size and by the structure of male and female genitalia. Ratio length: width of fore wing in brachypters = 1.56–1.9. Male pygofer as in Text-figs. 680, 681, ventral incision of pygofer as in Text-fig. 682, anal tube of male as in Text-fig. 683, genital style as in Text-fig. 684, aedeagus as in Text-figs. 685, 686. Venter of posterior part of female abdomen as in Text-fig. 687, genital scale as in Text-fig. 688. Length of brachypters 1.9–2.4 mm. I have not seen macropters of this species.

Distribution. Rare in Denmark, only found in NEJ: Tornby klit 23.VII.1967 by L. Trolle. In Sweden so far only found in Ög. (several localities), Vg., Kinnekulle, and in Upl. Not yet recorded from Norway. Rare in East Fennoscandia, only found in Al, Ab, Ta, LkE, and Le. – Austria, France, German D.R. and F.R., England, Netherlands, Poland, Romania, Switzerland, Latvia, Moldavia, Ukraine, Kazakhstan, Mongolia.

Text-figs. 680–688. *Ribautodelphax angulosus* (Ribaut). – 680: male pygofer from behind; 681: male pygofer from the right; 682: ventral margin of male pygofer from below; 683: male anal tube from behind; 684: left genital style; 685: aedeagus from the right; 686: aedeagus from the left; 687: caudal part of female abdomen from below; 688: genital scale from above. Scale: 0.25 mm for 687, 0.1 mm for the rest.

206

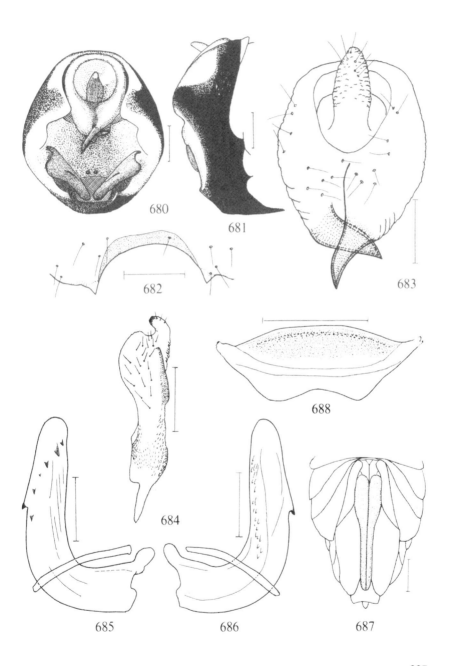

680

681

682

683

684

685

686

687

688

207

Biology. On dry slopes, belonging to the "Corynephoretum agrostidetosum aridae" (Marchand, 1953). "Bewohnt feuchte bis sumpfige Wiesen" (Wagner & Franz, 1961). On dry grassy places (Le Quesne, 1960). Our experience in Swedish conditions corroborates the statements of Marchand and Le Quesne rather than that of Wagner & Franz. We found adults in June, July, and August.

82. *Ribautodelphax pungens* (Ribaut, 1953)
   Text-figs. 689–692.

Calligypona pungens Ribaut, 1953: 247.

Resembling *collinus* and *angulosus*, differing by details in the male genitalia. Ratio length: width of fore wing in brachypters (one specimen measured) = 1.64. Male pygofer as in Text-figs. 689, 690, genital style as in Text-fig. 691, aedeagus as in Text-fig. 692. Length of brachypterous males (according to Le Quesne, 1960) 2.3–2.6 mm. I have not seen females nor macropters of this species.

Distribution. Sweden: Gtl. (R. Remane in litt.). – Not found in Denmark, Norway or East Fennoscandia. – Austria, Bohemia, Moravia, France, German D.R. and F.R., England, Italy, Netherlands, Poland.

Biology. Xerophilous character species of dry grass meadows (Schwoerbel, 1956). "Bewohner stark besonnter Trockenrasen und Felsenheiden" (Wagner & Franz, 1961). Stenotope species of dry grass meadows (Schiemenz, 1969). Hibernation in larval stage (Strübing in Müller, 1957, Schiemenz, 1969).

83. *Ribautodelphax pallens* (Stål, 1854)
   Text-figs. 693–702.

Delphax pallens Stål, 1854: 192.

Resembling *R. collinus*, differing by details in the structure of genitalia and by a smaller body. General colour of female sometimes darker, resembling *Hyledelphax elegantulus* ♀. Ratio length: width of fore wing in brachypters = 1.48–1.60. Male pygofer as in Text-figs. 693, 694, male anal tube as in Text-fig. 695, genital style as in Text-fig. 696, aedeagus (Text-figs. 697–699) evenly arched. Ventral aspect of posterior part of female abdomen as in Text-fig. 700, genital scale as in Text-figs. 701, 702. Length of brachypters 2–2.8 mm, of macropters 3.25–3.7 mm.

Distribution. Not in Denmark. – Fairly common in Sweden, found in Sm., Gtl., Ög., Vg., Sdm., Upl., Dlr., Nb., and Lu. Lpm. – Holgersen collected *R. pallens* in Norway, On: Vålåsjø, Dovre, and in HEn: Dalholen, Folldal, and HEn: Folldals Verk. – Common in East Fennoscandia, found in Al, Ab, N, St, Ta, Kb, Ok, Ks, LkW; Kr, Lr. – Austria, German F.R., England, Estonia, Latvia, Kaliningrad district, n. Russia.

208

Biology. In dryish fields but also, though in lower abundance, in moist sloping meadows and cultivated fields (Linnavuori, 1952). Adults in June and July.

84. *Ribautodelphax albostriatus* (Fieber, 1866)
Text-figs. 703–710.

Delphax albostriata Fieber, 1866b: 525.
Delphax distinguenda Kirschbaum, 1868: 23.

Resembling *collinus* but darker in colour, vertex and notum usually darker yellow-

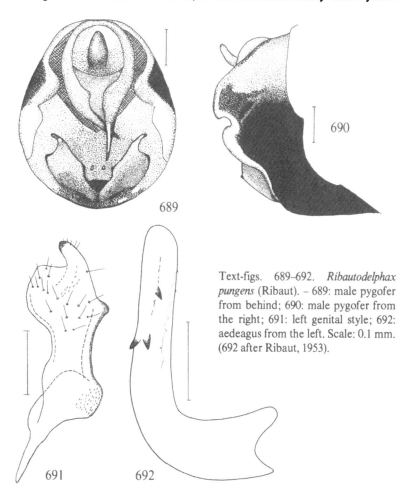

Text-figs. 689–692. *Ribautodelphax pungens* (Ribaut). – 689: male pygofer from behind; 690: male pygofer from the right; 691: left genital style; 692: aedeagus from the left. Scale: 0.1 mm. (692 after Ribaut, 1953).

209

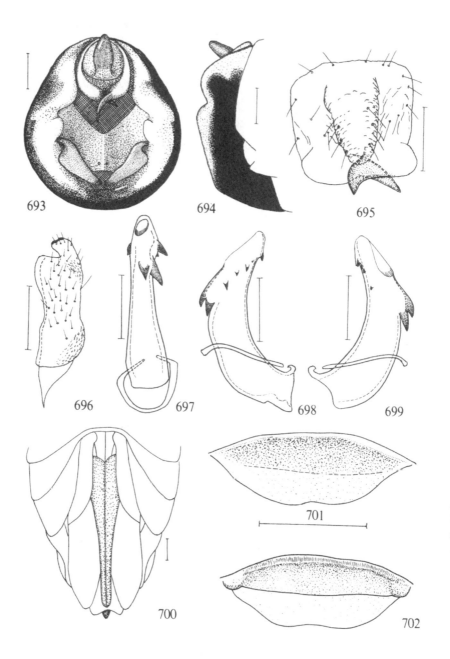

693

694

695

696

697

698

699

700

701

702

210

brownish, fore wings of brachypters transparent or semi-opaque, yellowish brown or dirty yellow with light margins, veins sometimes fuscous. Abdomen of male with some longitudinal rows of light spots. Abdomen of female laterally strongly dark-spotted, venter of body to a large extent dark. Light median band on vertex and thorax usually strongly marked, often continuing also on abdominal tergum. Ratio length: width of fore wings in brachypters = 1.49–1.72. Veins of fore wings in macropters fuscous or dark brownish. Male pygofer as in Text-figs. 703, 704, male anal tube as in Text-fig. 705, genital styles as in Text-fig. 706, aedeagus (Text-figs. 707, 708) of uniform thickness, evenly curved. Venter of caudal part of female abdomen as in Text-fig. 709, genital scale as in Text-fig. 710. Length of brachypters 2.4–3 mm, of macropters 3.6–4.3 mm.

Distribution: Not in Denmark. – Rare in the south of Sweden, common in the north (Bl., Sm., Gtl., Upl., Dlr., Hls., Hrj. – Nb., Ly. Lpm., Lu. Lpm., T. Lpm.). – Siebke (1874) recorded *Liburnia distinguenda* Kirschb. from Norway, AK: Oslo; Lindberg (1935) found *albostriatus* abundantly in Reg. arctica and subarctica in On: Dovre, Fokstua. H. Holgersen collected it in HEn: Folldals Verk. – Fairly common in East Fennoscandia (Ab, N, Ta, Oa, Sb, Kb, Ok, Ks, LkW, LkE; Vib, Kr, Lr). – Widespread in Europe, also in Cyprus, Tunisia, Caucasus, Kazakhstan, Kirghizia, m. Siberia, and Mongolia.

Biology. On sandy fields and sonny slopes (Kuntze, 1937). Listed among "meist stenotope Zikaden" of dry grass fields (Schwoerbel, 1956). Hibernation takes place in larval stage (Strübing in Müller, 1957, Schiemenz, 1969). In Sweden adults occur in May, June, July, and August. Macropters are not rare but less common than brachypters.

Family Achilidae

Antennal flagellum unsegmented. Head narrower than thorax. Median ocellus absent, lateral ocelli situated outside lateral carinae of frons. Pronotum short, hind border deeply concave. Second post-tarsal segment comparatively long. Wings always fully developed, no dimorphism. Clavus closed, veins distinct, not granulate, their common stem ending in apex of clavus. Fore wings widening distally of apex of clavus (Text-fig. 711). Anal area of hind wings (Text-fig. 712) not reticulate. Abdomen in adults without wax pores and basal processes. Male with pygofer horizontally flattened, paired medioventral process generally present, parameres large, complex. Female with ovipositor incomplete. In Fennoscandia one genus with two species.

Text-figs. 693–702. *Ribautodelphax pallens* (Stål). – 693: male pygofer from behind; 694: male pygofer from the right; 695: male anal tube from behind; 696: left genital style; 697: aedeagus, dorsal aspect; 698: aedeagus from the left; 699: aedeagus from the right; 700: caudal part of female abdomen from below; 701: genital scale from below; 702: genital scale from above. Scale: 0.25 mm for 700, 0.1 mm for the rest.

211

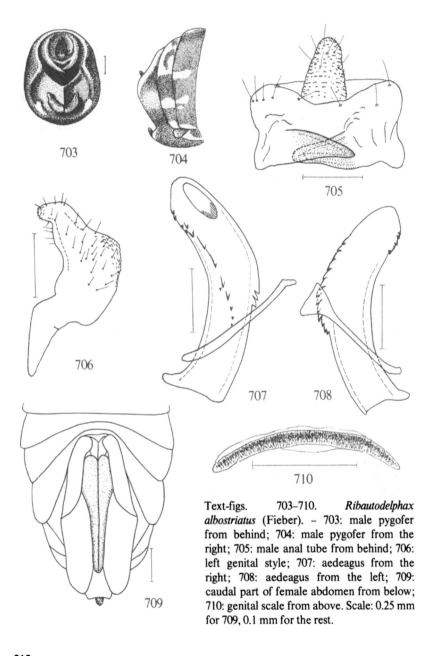

Text-figs. 703–710. *Ribautodelphax albostriatus* (Fieber). – 703: male pygofer from behind; 704: male pygofer from the right; 705: male anal tube from behind; 706: left genital style; 707: aedeagus from the right; 708: aedeagus from the left; 709: caudal part of female abdomen from below; 710: genital scale from above. Scale: 0.25 mm for 709, 0.1 mm for the rest.

Genus *Cixidia* Fieber, 1866

Cixidia Fieber, 1866a: 499.
Type-species: *Cixius confinis* Zetterstedt, 1838, by monotypy.
Epiptera Metcalf, 1922: 263.
Type-species: *Flata opaca* Say, 1830, by original designation.
Body oblong, depressed. Vertex depressed, protruding in front of eyes. Frons broadest in lower part, with one median carina, lateral margins carinate. Antennae short, first segment very short, ring-shaped, second segment globose or nearly so. Rostrum thin, apical segment prolonged, sometimes just reaching, sometimes reaching beyond apices of hind coxae. Pronotum medially about as long as vertex, fore border medially protruding, hind border concave, medially obtuse-angled. Pronotum with 3 carinae, lateral carinae caudally strongly diverging. Mesonotum with 3 carinae, these sometimes partly or entirely obsolescent. Fore wings at rest carried horizontally. Hind tibiae just beyond middle with a spine on outside, apically with several short pointed spines. First segment of hind tarsi prolonged.

Key to species of *Cixidia* (adults)

1 Vertex shorter than basal width. Dorsum and fore wings largely black-brown or dark brown, with numerous small whitish spots. Frons largely black-brown with some white markings   85. *confinis* (Zetterstedt)
– Vertex as long as basal width. Dorsum and fore wings largely yellowish brown. Frons largely yellow, apically fuscous   86. *lapponica* (Zetterstedt)

Key to 5th instar nymphs of *Cixidia*

1 Vertex anteriorly roundish. Lateral carinae of pronotum more or less parallel
                                                                    85. *confinis* (Zetterstedt)
– Vertex hexagonal, anteriorly blunt. Lateral carinae of pronotum caudally divergent
                                                                    86. *lapponica* (Zetterstedt).
(See Linnavuori, 1951: 62, Fig. 10).

85. *Cixidia confinis* (Zetterstedt, 1838)
Plate-fig. 21, text-figs. 711–719.

Cixius confinis Zetterstedt, 1838: 304.

Vertex pentagonal. Mesonotum with three carinae and between these traces of two additional keels. Side margins of frons convex, frons broadest below middle, as broad at basis as at apex. Black-brown. Clypeus, a narrow transverse band on frons, apex and base of genae, carinae of pronotum, anterior part of mesonotal carinae, and apex of

213

scutellum, yellowish white. Fore wings dark brown with more or less numerous, scattered, small, irregularly shaped whitish streaks and points. Fore wing membrane between veins finely transversely wrinkled. Venation of fore and hind wings as in Text-figs. 711, 712, male genital segment as in Text-figs. 713, 714, genital styles as in Text-figs. 715, 716; aedeagus long and thin, apex as in Text-figs. 717, 718. Apex of female abdomen as in Text-fig. 719. Length 5.3–7 mm.

Distribution. Not in Denmark, nor in Norway. – Very rare in Sweden. First found in Ly. Lpm.: Stensele by Dahlbom, later in Vg. by Gyllenhal and in Ög. by Wahlberg. Repeatedly found in Gotska Sandön by Mjöberg, Anton Jansson, and J. Jonasson. Lundblad collected *C. confinis* in Uppsala, "Fiby primeval forest". Bengt Ehnström found it in Dlr.: Nås, Säfsflotta, 21.V.1966. – Very rare also in East Fennoscandia. Found in Ab: Kimito, Dragsfjärd 13.VII.1920 by Wasastjerna, in Ta: Lammi 18.VI.1950 by Linnavuori, and in Sa: Luomäki by G. von Numers. – Outside Fennoscandia only found in Estonia.

Biology. In fissures in logs belonging to a demolished old wood building (Jansson, 1935: 59). "It was found that, in the limited part of the log infested by the *Cixidia*

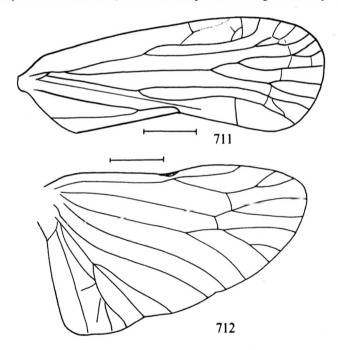

Text-figs. 711, 712. *Cixidia confinis* (Zetterstedt). – 711: right fore wing (venation only); 712: right hind wing. Scale: 1 mm.

214

specimens, the wood was still white and fairly firm, not disintegrated. Where the wood in the log was grey and loose the insect was not found. No growth of fungi in the wood was established". (Jansson, l. c.). "Larvae were found in the splintered wood of a storm-broken pine-stump inside the spruce forest on 6.X.1946. On 7.V.1947 parts of the trunk of this stump were brought home. During the summer of 1947 the following numbers of adults emerged: 1.VI.: 2, 2.VI.: 2, 3.VI.: 1, 4.VI.: 3, 5.VI.: 1, 9.VI.: 1, 30.VI.: 2. At a visit to the find place 9.VII.1947 six specimens were found sitting under the fairly loose bark of

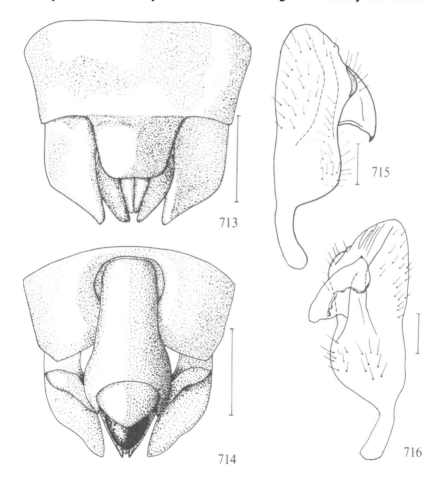

Text-figs. 713–716. *Cixidia confinis* (Zetterstedt). – 713: male genital segment from below; 714: male genital segment from above; 715: right genital style from outside; 716: right genital style from inside. Scale: 0.25 mm for 713 and 714, 0.1 mm for 715 and 716.

215

Plate-fig. 1. *Cixius cunicularius* (L.), × 6.

Plate-fig. 2. *Pentastiridius leporinus* (L.), × 6.

Plate-fig. 3. *Kelisia guttula* (Germ.) ♂, × 18.

Plate-fig. 4. *Stiroma bicarinata* (H.-S.) ♂ f. brach., × 15.

Plate-fig. 5. *Ditropis pteridis* (Spin.) ♂ f. brach., × 19.

Plate-fig. 6. *Eurysa lineata* (Perr.) ♀ f. brach., × 12.

Plate-fig. 7. *Delphacinus mesomelas* (Boh.), ♂ f. brach., × 20.

Plate-fig. 8. *Achorotile albosignata* (Dahlb.) ♀ f. brach., × 17.

Plate-fig. 9. *Euconomelus lepidus* (Boh.), left fore wing of macropterous specimen, × 24.

Plate-fig. 10. *Delphax crassicornis* (Panz.) ♂ f. macr., × 12.

Plate-fig. 11. Same, ♀ f. brach., × 12.

Plate-fig. 12. *Delphax pulchellus* (Curt.), left fore wing of macropterous specimen, × 12.

216

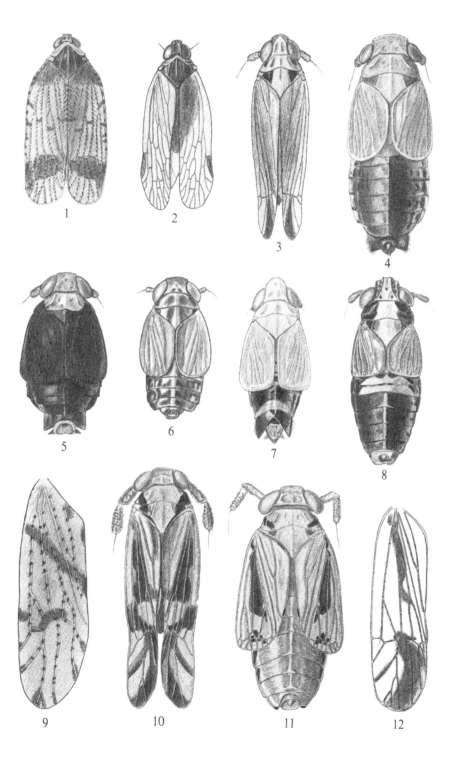

1   2   3   4

5   6   7   8

9   10   11   12

Plate-fig. 13. *Euides speciosa* (Boh.), ♂, × 11.

Plate-fig. 14. *Chloriona smaragdula* (Stål) ♂, × 12.

Plate-fig. 15. *Javesella pellucida* (F.) ♂ f. macr., × 15.

Plate-fig. 16. *Hyledelphax elegantulus* (Boh.) ♂ f. brach., × 18.

Plate-fig. 17. *Megamelus notula* (Germ.) ♂ f. brach., × 18.

Plate-fig. 18. *Dicranotropis hamata* (Boh.) ♂ f. brach., × 15.

Plate-fig. 19. *Oncodelphax pullulus* (Boh.) ♂ brach., × 26.

Plate-fig. 20. *Criomorphus albomarginatus* Curt. ♂ f. brach., × 16.

Plate-fig. 21. *Cixidia confinis* (Zett.), × 8.

Plate-fig. 22. *Cixidia lapponica* (Zett.), × 7.

Plate-fig. 23. *Issus muscaeformis* (Schrnk), × 7.

Plate-fig. 24. *Ommatidiotus dissimilis* (Fall.) ♀ f. macr., × 13.

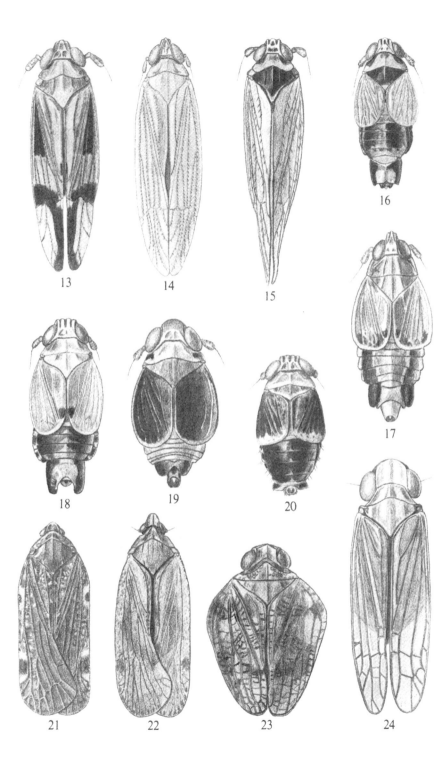

Plate-fig. 25. *Cixius nervosus* (L.) ♂, × 9.

Plate-fig. 26. *Ditropis pteridis* (Spin.) ♀ f. brach., × 14.

Plate-fig. 27. *Stenocranus minutus* (F.) ♂, × 14.

Plate-fig. 28. *Euconomelus lepidus* (Boh.) ♀ f. brach., × 20.

Plate-fig. 29. *Conomelus anceps* (Germ.) ♂ f. brach., × 18.

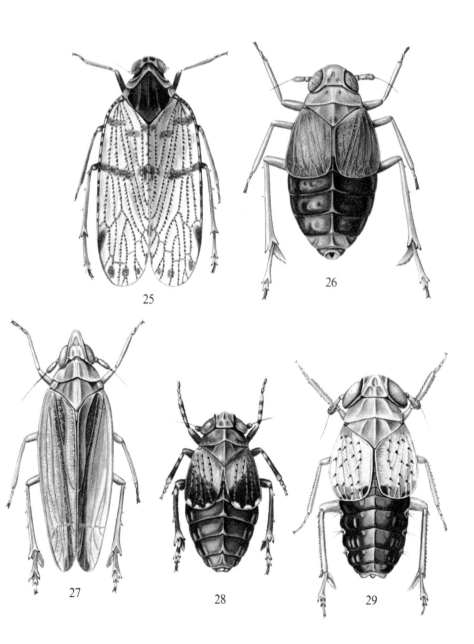

25

26

27

28

29

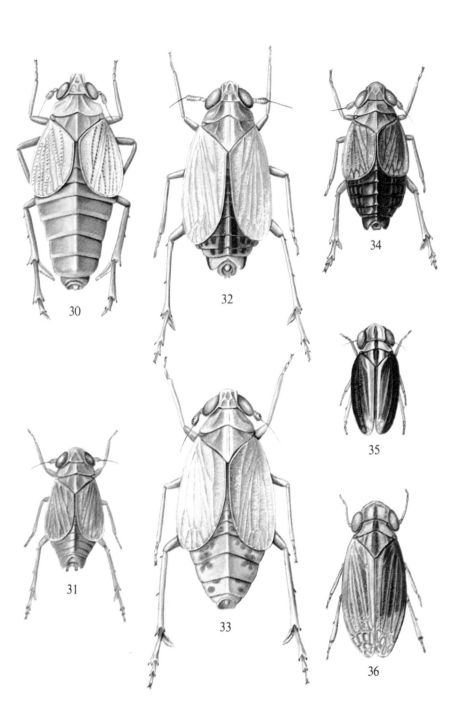

a part of the pine trunk having been broken away from the stump and lying on the ground" (Lundblad, 1950). "The larvae were found only in places in the trunk where the wood was rather hard, white and dry. They occurred in small colonies, about 10–15 larvae in each, ca. 2–5 cm under the surface. In the trunk was also a whitish *Poria* fungus" (Linnavuori, 1951). – The specimens collected by J. Jonasson were found in decayed pine firewood indoors. von Numers found his specimens under pine-bark, in wood fissures, with mycelium of a fungus. – Apparently hibernation can take place in the larval stage. Adults in our collection are dated 20.V., 21.V., and 10.VIII.

86. *Cixidia lapponica* (Zetterstedt, 1838)
Plate-fig. 22.

Cixius lapponicus Zetterstedt, 1838: 304.
Cixidia confinis ♀ Jansson, 1935: 60, Fig.

Vertex elongate. Frons upwards narrowing, carinae sharp. Mesonotum with three part-ly indistinct carinae. Body oblong, depressed, brownish yellow. Face, apex of scutellum, and legs whitish yellow; fore wings semi-transparent, finely dark-mottled, with dark veins. Frons above with a broad brownish transverse band. Vertex apically with three black points. Pronotum dirty yellow with lighter dotting. Thorax ventrally yellowish, with a dark lateral longitudinal band on each side. Mesonotum brownish with numerous lighter dots. Abdomen black-brownish. Length with wings 7–8.5 mm.

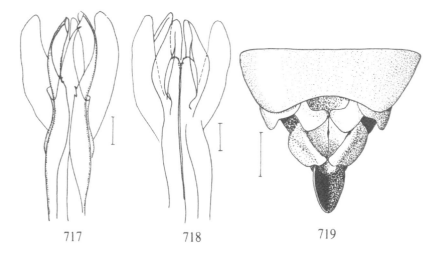

717   718   719

Text-figs. 717–719. *Cixidia confinis* (Zetterstedt). – 717: apex of aedeagus, dorsal aspect; 718: apex of aedeagus, ventral aspect; 719: apex of female abdomen from below. Scale: 0.25 mm for 719, 0.1 mm for the rest.

Distribution. So far not found in Denmark, nor in Norway. – Very rare in Sweden: Gotska sandön (A. Jansson); Ög. (Wahlberg); Dlr. (Boheman), Dlr: Hamra (A. Jansson), Leksand, Sången 20.VI.1937 (Tord Tjeder), Nås, Säfsflotta 20.V.1966 (B. Ehnström); "Lp. in." (Boheman); Ås. Lpm.: Långseleberg (Dahlbom); Ly. Lpm.: Stensele (Dahlbom), Lycksele, Granhöjden 17.VII.1946 (Forsslund). – Very rare also in East Fennoscandia. Found in St: Yläne (C. Sahlberg); Ka: Lammi (Linnavuori), Hattula 28.VII.1950 (Matti Nuorteva); LkW: Muonioniska, Kätkäsuando 2.IX.1867 (J. Sahlberg); Kr: ?Petrosavodsk (Günther), Lac. Hirvas (Axelson). – N. Russia, m. and e. Siberia.

Biology. Apparently similar to that of *Cixidia confinis*. Not only Dahlbom (Zetterstedt, l. c., 305), but also Jansson (1935), Linnavuori (1951), and Ehnström (personal communication) found both species together.

Family Issidae

Body robust. Ocelli absent. Pronotum short, especially behind eyes, hind border straight or faintly concave. Mesonotum short. Wing monomorphous or rarely polymorphous species. In monomorphous species and in the brachypterous form of polymorphous species the fore wings are leathery, often very convex. Clavus not granulate. Hind tibiae usually with one or a few spines on outside. Second segment of hind tarsi with a spine on each side. Male pygofer with aedeagus ventrally with a pair of styles varying in shape, connected with aedeagus by a connective. Structure of aedeagus usually very complicated. Gonapophyses of female short and broad, robust. On woody and herbaceous plants, ecology of larvae same as that of adults. In Denmark and Fennoscandia two genera.

Key to genera of Issidae

1   Head with eyes a little broader than pronotum. Body and fore wings comparatively narrow. Wing polymorphous. Length of macropters about 5 mm, brachypters shorter. Hind tibiae on outside with 1 spine   *Ommatidiotus* Spinola (p. 220)
–   Head with eyes narrower than pronotum. Body short and broad. Wing monomorphous, fore wings very broad, sides strongly subangularly dilated before middle. Length (in our species) over 5 mm. Hind tibiae with 2 spines on outer side
                                                                                      Issus Fabricius (p. 218)

Genus *Issus* Fabricius, 1803

Issus Fabricius, 1803: 99.
    Type-species: *Cercopis coleoptrata* Fabricius, 1781, by subsequent designation.

218

Vertex pentagonal. Frons below upper limit with a transverse carina. Eyes large. Pronotum with a median carina, lateral carinae absent. Scutellum with three carinae. Fore wings 1/3 from basis strongly dilated, distally narrowing. Hind tibiae on outside with two strong spines. In Denmark and Fennoscandia one species.

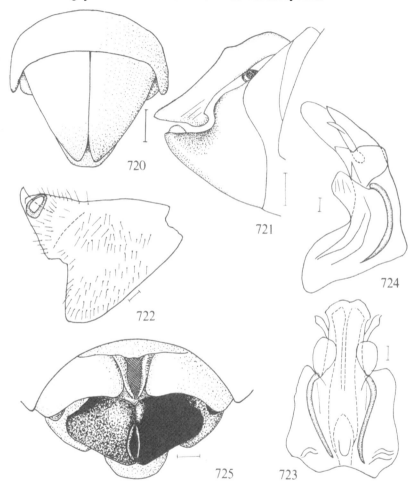

Text-figs. 720–725. *Issus muscaeformis* (Schrank). – 720: apex of male abdomen from below; 721: apex of male abdomen from the right; 722: right genital style from outside; 723: aedeagus from above (specimen from Dalmatia, Ragusa); 724: aedeagus from the right (same specimen); 725: apex of female abdomen from below. Scale: 0.25 mm for 720, 721, and 725, 0.1 mm for the rest.

219

Key to species of *Issus*

1 Frons below with a broad pale transverse band occupying lower 1/3. Fore wings with two indistinctly limited fuscous transverse bands (Plate-fig. 23). Longitudinal veins of fore wings usually distinct throughout   87. *muscaeformis* (Schrank)
- Frons without a distinct pale transverse band. Fore wings dirty whitish with partly fuscous veins and one or two dark spots 1/3 from apex, in apical third often with many irregular veinlets making longitudinal veins partly indistinct. In England, France, Germany etc.   *coleoptratus* (Fabricius)

87. *Issus muscaeformis* (Schrank, 1781)
Plate-fig. 23, text-figs. 720–725.

Cicada muscaeformis Schrank, 1781: 253.
Issus frontalis Fieber, 1876: 264.

Greyish yellow, sometimes more greenish, colour pattern much varying. Frons laterally black with yellowish granules, lower 1/3 pale. Genae pale, postclypeus fuscous, laterally darker. Deflexed lateral part of pronotum pale below. Femora partly fuscous, fore and median tibiae basally and apically dark and with a broad dark zone proximally of middle. Veins of fore wings largely fuscous, except in an irregular transverse light band somewhat proximally of middle. Apex of male abdomen as in Text-figs. 720, 721, genital style as in Text-fig. 722, aedeagus as in Text-figs. 723, 724. Apex of female abdomen as in Text-fig. 725. Length with wings 5.5–7 mm.

Distribution. Denmark: scarce in central Jutland (EJ, WJ), also found in B: Helligdomsklipperne 6.XI.1972 (L. Trolle). – Scarce in the south of Sweden: Sk., Bl., Hall., Sm., Öl., Sdm. – In Norway found in AAy: Risør (Warloe); VE: Sandar, Arø 29.VII.1969 (L. Greve); HOi: Strandebarm Bakke, "Eikenes" 1.VII.1970 (L. Greve). – So far not found in East Fennoscandia. – Austria, Bulgaria, Czechoslovakia, France, German F.R., Greece, Hungary, Italy, Netherlands, Poland, Romania, Armenia, Georgia, Ukraine, Yugoslavia.

Biology. On *Quercus* in sun-exposed sites. Adults in June and July. I found larvae in April and in August, so hibernation probably takes place in larval stages.

Genus *Ommatidiotus* Spinola, 1839

Ommatidiotus Spinola, 1839: 365.
Type-species: *Issus dissimilis* Fallén, 1806, by monotypy.

Vertex pentagonal, anteriorly angular or rounded. Eyes large. Frons about as long as broad, sides convex, with 3 carinae, lateral carinae parallel with lateral margins. Clypeus with a median carina. Pronotum trapezoidal, shorter than vertex. Mesonotum large, with 3 straight carinae, the median one indistinct. Hind tibiae on outside with one

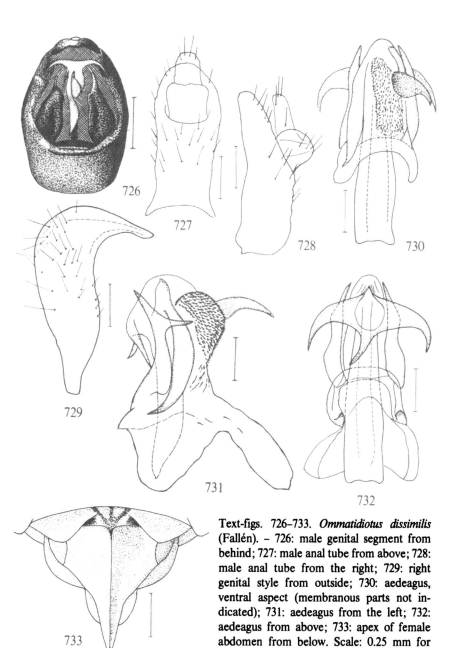

Text-figs. 726–733. *Ommatidiotus dissimilis* (Fallén). – 726: male genital segment from behind; 727: male anal tube from above; 728: male anal tube from the right; 729: right genital style from outside; 730: aedeagus, ventral aspect (membranous parts not indicated); 731: aedeagus from the left; 732: aedeagus from above; 733: apex of female abdomen from below. Scale: 0.25 mm for 726, 0.1 mm for the rest.

strong spine. Wing polymorphous species. Fore wings of brachypters a little longer, those of macropters much longer than abdomen. 1st and 2nd segments of hind tarsi thickened. In Denmark and Fennoscandia one species.

88. *Ommatidiotus dissimilis* (Fallén, 1806)
Plate-figs. 24, 35, 36, text-figs. 726–733.

Issus dissimilis Fallén, 1806: 123.

Body largely black, with a fine erect pilosity. Wing polymorphous. Fore wings of brachypters (both sexes) leathery, just longer than abdomen. Frons of brachypterous male entirely black, that of brachypterous female black with yellowish carinae. Dorsum straw-coloured. Claval commissure orange-coloured, an orange-coloured longitudinal band continuing forwards on thorax and head. This band is often bordered by a darker line and is sometimes entirely dark brownish. In the male (Plate-fig. 35), about half surface of the fore wing is occupied by a broad black longitudinal band along costal border, rest of fore wing whitish with orange-coloured longitudinal bands in the cells. In the brachypterous female (Plate-fig. 36), as well as in a rare male variety, the wing is entirely whitish with orange longitudinal bands in the cells. Macropters (Plate-fig. 24) have hind wings fully developed, fore wings extending behind apex of abdomen by about 1/3 of their length. Specimens intermediary in wing length do also exist. The fore wings of macropterous and intermediary individuals are in major part colourless and transparent, basally dark-shaded, with venation in apical part irregularly ramified. Male pygofer as in Text-fig. 726, male anal tube as in Text-figs. 727, 728, genital style as in Text-fig. 729, aedeagus as in Text-figs. 730–732. Apex of female abdomen as in Text-fig. 733. Overall length of brachypters 2.7–4.75, of macropters 4.8–5.6 mm.

Distribution. Denmark: common in Jutland (SJ, EJ, WJ, NWJ, NEJ). – Comparatively common, locally abundant in the south of Sweden, Sk. – Dlr. (not in Öl., Gtl. and G. Sand.). – Norway: Ø: Øymark, Vinsknatten 18.VIII.1960 (Holgersen), Ø: Holon, Degernes 25.VIII.1960 (Holgersen). – Comparatively rare in East Fennoscandia, found in Ab, N, St. Sa, and Kb; Vib and Kr. – Not in Great Britain, nor in Mediterranean countries except n. Italy, otherwise widespread in Europe, also in Altai, Kazakhstan, Kirghizia, w. and m. Siberia, and Mongolia.

Biology. In bogs, on *Eriophorum vaginatum* (Kuntze, 1937). Hibernation takes place in the egg stage (Müller, 1957, Remane, 1958). Adults in July–September.

Printed in the United States
By Bookmasters